Statistics and the MINITAB®
Windows Program:

COMPUTER LAB
EXPERIMENTS

LLOYD R. JAISINGH
ROBERT J. LINDAHL

Morehead State University

Saunders College Publishing

HARCOURT BRACE COLLEGE PUBLISHERS

Fort Worth   Philadelphia   San Diego   New York   Orlando   Austin
San Antonio   Toronto   Montreal   London   Sydney   Tokyo

MINITAB® Statistical Software is a registered trademark of
MINITAB, Inc.

Windows™ is a trademark of Microsoft Corp.

Printed in the United States of America.

Jaisingh & Lindahl; Lab Manual to accompany <u>Statistics and the
Minitab Windows Program: Computer Lab Experiments, First
Edition.</u> Hawley

ISBN 0-03-017733-2

567 021 987654321

PREFACE

Because of the rapid changes in technology, the study of statistics must undergo significant changes. Our teaching methods must be reevaluated to incorporate the technology to help students understand the probability and statistical concepts that are presented to them. The enclosed statistical labs are designed for any elementary statistics course. These labs use the **MINITAB® for Windows** software to alleviate much of the computational drudgery, thus enabling the students to concentrate on the discovery, application, and reinforcement of the concepts presented in their lectures.

The primary goal of these labs is to implement the use of technology in the teaching of any elementary statistics course. It is hoped that the use of these labs will encourage students to learn through discovery in a relaxed environment.

This manual contains sixteen labs (0 - 15) and cover a wide range of topics found in any elementary statistics course. These labs present topics in sequence as would be found in most elementary statistics texts. However, the instructor can decide on the appropriate order for his/her course.

The **MINITAB for Windows** software was chosen because of the combination of its power and ease with which the students can learn to use it. Also, since **MINITAB** software is used in about 70% of all fortune 500 companies, it is important that students have hands-on experience with it. Each lab includes a review section of the appropriate topics covered for the particular lab. Detailed step-by-step instructions are given in the labs as to which commands to use to obtain the desired results. At the end of each lab there is a *DATA SHEET* which the students will be required to work and turn in as assignments, when appropriate. In some instances, assignments are included which can be assigned as group exploratory projects. Keep in mind, these labs are intended for discovery and should not be used as a tutorial guide for **MINITAB**. Moreover, once a student has completed these labs, they most certainly will be very knowledgeable about the **MINITAB** software.

It would be beneficial to the students if they work in pairs in the computer lab since the discussions that occur as they work will help them to understand the concepts with which they are dealing.

DEDICATION

We would like to dedicate this work to our families.

STATISTICS and the MINITAB® WINDOWS PROGRAM: *LAB EXPERIMENTS WITH THE COMPUTER*

TABLE OF CONTENTS

STATISTICS LAB # 0

INTRODUCTION TO MINITAB

PURPOSE - explain MINITAB commands relating to

1. entering and editing data in the **Data** window spreadsheet
2. saving, loading, and printing a **worksheet**
3. the **Menu** bar and **Dialog** boxes
4. saving and printing the contents of various windows
5. accessing the **Help** files
6. the **Session**, **Info**, and **History** windows

1. GETTING STARTED WITH MINITAB

After turning on the computer and entering *Windows®*, locate the **MINITAB for Windows** icon and double click on it with the mouse. Figure *Minitab 0.1* shows a picture of what may now appear on the screen.

Minitab 0.1

1

The **Data** window is in the form of a spreadsheet, with 100,000 cells available as the default worksheet setting for the full version of **MINITAB**. The student version of **MINITAB** is limited to 3,500 cells. The default cell width is eight characters.

The columns are listed as C1, C2, C3, ... and are available for data entry. Each single set of data is entered in one of these columns. When data is entered in one or more columns, **MINITAB** commands are available in order to generate statistical information by selecting from the menu bar at the top of the screen.

Moving around in the **Data** window is accomplished by pressing any of the four arrow keys or by clicking on a cell with the mouse. The size of a window can be changed by a click and drag of the window border, or by clicking on the up or down arrow located at the top right. Window sizing can also be accomplished by clicking on the right or bottom slide bar.

Values can be entered in the **Data** window by one of three methods:
(a) typing them in the cells;
(b) loading a worksheet file;
(c) having **MINITAB** generate the values after selecting a menu option.

Example 1: The following data values were obtained from a survey of 10 households asking for an estimate of their average monthly long distance telephone bill:

Household No.	1	2	3	4	5	6	7	8	9	10
Monthly cost	$15	$21	$8	$23	$5	$36	$18	$6	$16	$11

In the **Data** window, enter the monthly cost values in column C1. After each value is entered, press the down arrow key. Below the C1 heading is a cell to place a column label with up to eight characters. Place a label of *Mo. Cost*. In figure *Minitab 0.2*, the **Data** window has been adjusted so as to reveal the **Session** window. The **Info** and **History** window icons are shown at the bottom of the display.

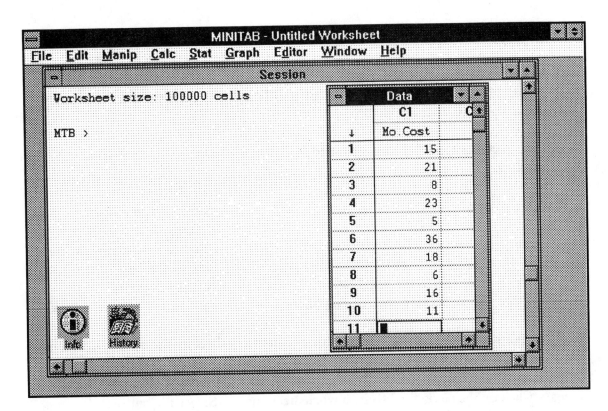

Minitab 0.2

2. MINITAB MENU BAR OPTIONS

The **menu bar** at the top of the worksheet window has nine options, namely

File	Edit	Manip	Calc	Stat	Window	Graph	Editor	Help

Here is a brief summary of some of the main features for each of these options:

(a) **File** - load, save, and print various windows and the worksheet. Also, values of columns and constants can be displayed in the **Session** window.

(b) **Edit** - copy, paste, and delete blocks of column or row values.

(c) **Manip** - modify and convert data columns; erase or copy whole columns and rows of current data values.

(d) **Calc** - operate on columns, rows, and constants; generate random data.

(e) **Stat** - produce various descriptive and inferential statistics from the current data.

(f) **Window** - to move from one window to another window; delete a graph window.

(g) **Graph** - create a graphical display of current data such as a histogram, bar graph, or scatter plot. There are two types of graphical displays, namely a character graph and a high resolution graph.

A character graph (such as a stem-and-leaf) will be displayed in the **Session** window.

A high resolution graph will be displayed in a newly created **Graph** window.

(h) **Editor** - modify cell format and cell size.

(i) **Help** - access the various help files which explain **MINITAB** commands. Help files can also be accessed by clicking on the question mark button in a dialog box.

To demonstrate how **File** can be used to save the worksheet in *Example 1*, click on **File→Save Worksheet As**. Interpret the arrow → to mean that **File** is first selected, followed by **Save Worksheet As**. The dialog window titled **Save Worksheet As** will appear. Click on the circle for **Minitab worksheet** and the **Select file** button.

The **Save Worksheet As** dialog box in figure *Minitab 0.3*.

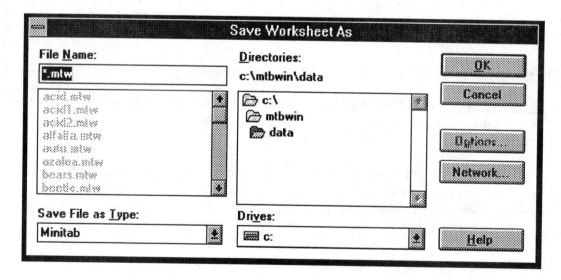

Minitab 0.3

In the **File Name** box will appear ***.mtw**. Click in this box and type in the file name *tel_cost.mtw*. The file extension **.mtw** is needed for each saved worksheet. To exit from **MINITAB**, select **File→Exit**.

Any saved worksheet can be retrieved by selecting **File→Open Worksheet→Select Worksheet**. A list of files having the **.mtw** can now be viewed. Click on the worksheet name to be retrieved followed by **OK,** or simply double click on this file name. The previously saved constants and variables will now reappear in the **Data** window.

Suppose that one is interested in finding the mean of the 10 monthly long distance costs. One way to accomplish this is through **Calc→Mathematical Expressions**. Observe how the boxes in the **Mathematical Expressions** dialog box are clicked and filled in as in figure *Minitab 0.4*.

	Mathematical Expressions

C1 Mo.Cost	**Variable (new or modified):**	k1
	Row number (optional):	
	Expression:	
	Sum(c1)/10	

Type an expression using:

Count	Min	Absolute	Sin	+ - * / ** ( )
N	Max	Round	Cos	= <> < <= > >=
Nmiss	SSQ	Signs	Tan	AND OR NOT
Sum	Sqrt	Loge	Asin	
Mean	Sort	Logten	Acos	Nscores
Stdev	Rank	Expo	Atan	Parsums
Median	Lag	Antilog		Parproducts

Select   Help   OK   Cancel

Minitab 0.4

The value of the mean is stored in constant K1 and the **Session** window displays the equation, but not value, relating to it. To find the value of any variable (constant, column, or matrix), click on **File→Display Data**. Type *K1* in the dialog box and click on **OK**. The mean of the 10 monthly costs is displayed in the **Session** window as 19.8.

Data columns are expressed in **MINITAB** with C1, C2, C3, ... Constants are expressed as K1, K2, K3, ..., and matrices as M1, M2, M3, ...

A second method of obtaining the mean, as well as several additional statistics related to a data column, is to select **Stat→Basic Statistics→Descriptive Statistics**.

3. HIGH RESOLUTION GRAPHICAL DISPLAYS OF DATA

One of the most important advancements made in **MINITAB for Windows** is the ability to display high resolution graphs of data. To demonstrate a **MINITAB** display of a histogram of a data set, consider the following example. First, erase the previous entries in column C1 by clicking on C1 with the mouse and pressing the **Delete** key.

Example 2: The at-rest pulse rates, over a one minute period, of 50 adult males is recorded in the following table.

75	71	66	58	77	82	69	72	65	77	70	75	72
74	63	79	71	74	85	49	58	73	71	66	80	75
69	66	78	65	69	77	73	74	81	80	75	58	
77	74	76	70	68	67	74	78	70	72	74	77	

Enter these 50 values in column C1 and label the column as *Pulse*. A no-frills histogram can now be constructed by selecting **Graph→Histogram**. Fill the **Histogram** dialog box as shown in figure ***Minitab 0.5***.

Minitab 0.5

6

After **OK** is clicked, a **Graph** window is created by **MINITAB** containing a display of a histogram of pulse rates, as shown in figure *Minitab 0.6*.

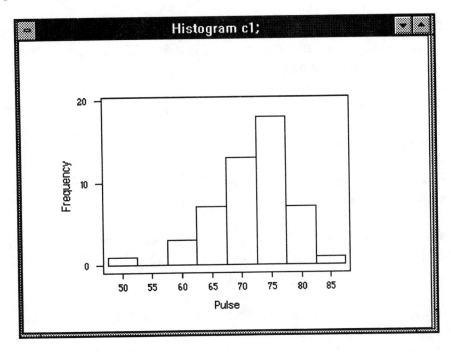

Minitab 0.6: *Histogram of 50 Male Pulse Rates*

Other types of histograms can be displayed by placing a title on the graph, modifying the horizontal and vertical titles, and changing the position and number of bars. These modifications will be addressed in later lab sessions.

A **Graph** window can be saved by clicking on **File→Save Window As**. A dialog box will appear as displayed earlier in figure *Minitab 0.3*, with the extension ***.mgf** now appearing in the box below **File Name**. As an example, the **Graph** window containing the histogram from *Example 2* could be saved by typing *heart.mgf*.

A previously saved graph file can be loaded back into a worksheet by selecting **File→Open Graph**. The dialog box will reveal a listing of all the graph files (with file extension **.mgf**) in the **MINITAB** directory.

One way to erase a **Graph** window from a worksheet is to make the window active, click on the top left square, and select **Discard Graph**.

4. THE SESSION, HISTORY, AND INFO WINDOWS

Most of the discussion at this point has focused on the **Data** window. The other three windows which are always present when engaged in a **MINITAB** work session are the **Session**, **History**, and **Info** windows.

The **Session** window provides a running display of actions taken in a work session.

One can type in a command statement in the **Session** window, instead of clicking on the equivalent command on the menu bar. This is the process used for all earlier versions of **MINITAB**. For instance, the *Pulse* label on column C1 in *Example 2* could have been entered in the **Session** window by typing *NAME C1 'Pulse'*.

The **History** window displays all session commands, without any of the output present from the **Session** window. It is useful to review the contents in this window and obtain a printout of actions taken during the work session.

The **Info** window provides a display of all columns, matrices, and stored constants which have been created in the work session.

5. PRINTING THE CONTENTS OF A WINDOW

In order to obtain a printout of the information provided in a window, it is first necessary to set up a default printer in the **Program Manager** of *Microsoft Windows*.

When the menu option **File→Print Window** is selected, a dialog box will appear containing the default printer name and the three options of **All**, **Selection**, and **Pages**.

If only a part of the window is to be printed, the mouse must be clicked and dragged in order to highlight the desired text or cells.

For a printout of the information stored in the **Data window**, it is *very important* to highlight the desired cell block so as not to print out all cells in the spreadsheet!

A printout of the column C1 pulse readings of the 50 males in *Example 2* is produced after making the appropriate choices in the dialog box shown in figure *Minitab 0.7*.

```
┌─────────────────────────────────────────────────────────┐
│ ▬        Data Window Output Options                       │
├─────────────────────────────────────────────────────────┤
│                                                           │
│   ☒ Print Row Labels                    ┌──────────┐     │
│                                         │    OK    │     │
│   ☒ Print Column Labels (i.e. C1)       └──────────┘     │
│                                         ┌──────────┐     │
│   ☒ Print Column Names                  │  Cancel  │     │
│                                         └──────────┘     │
│   ☒ Print Grid Lines                                     │
│                                                           │
│   Column Names and Labels:                               │
│       ○ Left Justified                                   │
│       ○ Centered                                         │
│       ○ Right Justified                                  │
│       ◉ Numeric Right Justified; Alpha Left Justified    │
│                                                           │
│   Title:                                                  │
│  ┌──────────────────────────────────────────────────┐   │
│  │ Histogram of 50 Adult Male Pulse Rates           │   │
│  └──────────────────────────────────────────────────┘   │
└─────────────────────────────────────────────────────────┘
```

Minitab 0.7

9

NOTES

LAB #0: *DATA SHEET*

Name: _____   *Date:* _____

Course #: _____   *Instructor:* _____

These exercises are designed primarily to acquaint the student with various commands in **MINITAB for Windows**.

1. After you have entered **MINITAB**, type the 10 long distance monthly costs and *Mo. Cost* heading in column C1 as displayed in *Minitab 0.2*.

Household No.	1	2	3	4	5	6	7	8	9	10
Monthly Cost	$15	$21	$8	$23	$5	$36	$18	$6	$16	$11

(a) Determine the mean (average) cost and median (middle) cost of these 10 values by clicking on **Stat→Basic Statistics→ Descriptive Statistics**.   Perform the calculations of these two values by using their basic definitions and a calculator.

Mean = _____.   Median = _____.

Definitions used of mean and median:

The estimated average number of long-distance calls made each month for the 10 households is found in the following table:

Household No.	1	2	3	4	5	6	7	8	9	10
No. of Calls	16	12	15	25	5	20	6	10	30	20

Place a label of *No. Calls* for column C2 and type in the 10 values from the table.

(b) Determine the mean number of calls made in the 10 households by selecting **Calc→ Mathematical Expressions** and proceeding as shown in figure *Minitab 0.4*.

Mean = _____.

(c) Place a label of *Call Ave* on column C3 and select **Calc→Mathematical Expressions** (see figure *Minitab 0.4*). Fill in the **Variable** and **Expression:** boxes so that column C3 will now contain the 10 ratios C1/C2. What do the numbers in column C3 represent?

(d) Highlight the data in columns C1, C2, and C3 by clicking and dragging with the mouse. Then select **Files→Print Window** to print out the contents of the **Data** window. In the first dialog box you encounter, type in an appropriate title and check the boxes as indicated in figure *Minitab 0.7*.

2. Enter the 50 adult male pulse rates from *Example 2*. First, erase the data values in columns C1-C3 by highlighting the appropriate cells and pressing the **Delete** key.

75	71	66	58	77	82	69	72	65	77	70	75	72
74	63	79	71	74	85	49	58	73	71	66	80	75
69	66	78	65	69	77	73	74	81	80	75	58	
77	74	76	70	68	67	74	78	70	72	74	77	

(a) Use the menu options **Calc→Column Statistics** to find the mean pulse rate of the 50 adult males. In the dialog box, click on the **Mean** circle, enter *C1* in the **Input variable** box, and store the result in constant K1. Select **File→Display Data** and type *K1* in the dialog box in order to see the mean displayed in the **Session** window.

Mean pulse rate = _____.

(b) The menu option **Manip→Sort** enables one to sort a column of data values from smallest to largest. Place the label *Sort* on column C2. After selecting **Manip→ Sort**, type *C1* in the **Sort column:** box, *C2* in the **Store sorted column(s) in:** box, and *C1* in the **Sort by column box**. Use column C2 to find the percentage of pulse rates that are less than 70.

Percentage less than 70: _____.

(c) One important concept in statistics is the **deviation from the mean** which is the difference between a data value and the mean of all data values. The sum of the deviations is always zero.

Place a label of *Mean Dev* on column C3 and select **Calc→Mathematical Expressions**. In the dialog box, type *C2* in the **Variable:** box and *C1 - K1* in the **Expression:** box. Then select **Calc→Column Statistics** and fill in the dialog box so as to form the sum of all the deviations from the mean.

Sum of the deviations from the mean = _____

Is this what you would expect? Explain.

(d) Select **Graph→Histogram** (see figures *Minitab 0.5* and *Minitab 0.6*) to display a vertical projection graph of the 50 pulse rates. To do this, in the dialog box select **Display→Project**. Provide a printout of this **Graph** window. Describe the shape of this histogram.

(e) Save the **Graph** window created in (d) as *Heart.mgf* and save the worksheet as *Heart.mtw*. Explain the steps taken in order to accomplish this.

(f) Provide a printout of the three columns from the **Data** window. Be sure to highlight the block of cells from these columns! Place a proper title on the printout sheet. Provide printouts of the **Session**, **History**, and **Info** windows.

STATISTICS LAB # 1

GRAPHICAL DESCRIPTIVE METHODS AND NUMERICAL DESCRIPTIVE MEASURES

PURPOSE - to use MINITAB to

1. **construct** frequency tables
2. **construct** line (projection) graphs, bar graphs, frequency polygons, relative frequency graphs, pie charts
3. **construct** histograms for grouped data
4. **construct** stem-and-leaf plots
5. find the *sample* mean, standard deviation, variance, median, first quartile, third quartile, and the mode for ungrouped univariate data

BACKGROUND INFORMATION

1. A **frequency distribution** for *ungrouped* data is a table that gives each of the distinct values in the data set and their corresponding number of occurrences (frequency).

2. The sum of the frequencies is equal to the total number of values in the data set.

3. A **line (projection) graph** plots the successive values on a horizontal axis and indicates the corresponding frequencies by the height of each vertical line segment.

4. A **bar graph** or **bar chart** (for **qualitative** data) uses vertical or horizontal bars to represent the frequencies of the individual categories in the data set.

5. A **frequency polygon** for *ungrouped* data is a graph that displays the frequencies for the individual data values as points and then using straight line segments to connect the plotted points.

15

6. A **relative frequency graph** for *ungrouped* data is a graph that uses the relative frequencies for the individual data values along the vertical axis with the individual values along the horizontal axis. If the graph displays the relative frequencies for the individual data values as points, a **relative frequency polygon** is obtained by connecting these points with straight line segments.

7. A **pie chart** is often used to plot relative frequencies or percentages when the data are **qualitative**. A circle is constructed and divided up into sectors whose areas are proportional to the relative frequencies.

8. A **histogram** is a graph in which the data are divided into classes, with the classes plotted along the horizontal axis. Rectangles are constructed for these classes with the rectangles placed adjacent to each other. The vertical axis of a histogram can represent either the class frequency or relative class frequency. A histogram is primarily constructed for **quantitative** data.

9. **Stem-and-leaf** plots are data plots that use the most significant part of the data value as the stem and the least significant part of the data value as the leaf to form groups or classes.

10. A **scatter diagram** or **scatter plot** is a two-dimensional rectangular plot of a set of ordered pairs (paired values).

11. The **sample mean** $\bar{x}$ is the sum of the data values in the sample divided by the sample size, i.e.,

$$\bar{x} = \frac{\sum_{i=1}^{n} x_i}{n},$$ where n is the sample size and x_i are the sample values.

12. The **sample variance**, s^2, is given by $s^2 = \dfrac{\sum_{i=1}^{n} (x_i - \bar{x})^2}{n-1}$. A simpler equivalent formula is $s^2 = \dfrac{\sum_{i=1}^{n} x_i^2 - n\bar{x}^2}{n-1}$.

13. The **sample standard deviation** s is the positive square root of the sample variance, i.e., $s = +\sqrt{s^2}$.

14. The **sample median** for a set of values is that numerical value such that at most 50% of the values are smaller than this number (median) and at most 50% of the data set are larger than this value (median) *when the data values are ordered*.

15. The **sample mode** is that value occurring most often in the data set.

16. The ***k*th percentile** of an ***ordered*** data set, P_k, is that numerical value such that *at most k* % of the values are smaller than P_k, and *no more than* $(100 - k)$ % are larger than P_k.

17. The **first quartile** of a data set Q_1 is the 25th percentile.

18. The **third quartile** Q_3 is the 75th percentile.

19. The **interquartile range** is defined to be the difference $(Q_3 - Q_1)$.

20. A data point is considered an outlier if it is 1.5 times the interquartile range above Q_3 or 1.5 times the interquartile range below Q_1.

PROCEDURES

First, load the **MINITAB** (windows version) software as in *Lab #0*.

NOTE: The procedures presented in these labs may not be the only way to achieve the end results. Also, whenever graphs are presented, only the MINITAB graphics features are used.

1. CONSTRUCTING FREQUENCY TABLES

Example 1: Use **MINITAB** to construct frequency, cumulative frequency, relative frequency and cumulative relative frequency tables for the variable *HEIGHT* from the *trees.mtw* data worksheet. This worksheet comes with the **MINITAB** program.

To use the **trees.mtw** worksheet, you first need to open it. Select **File→Open Worksheet**. The **Open Worksheet** dialog box will be displayed. Use the mouse to scroll down the box below the **File Name:** box to select the file named *trees.mtw*. It will then be listed in the **File Name:** text box. The **Open Worksheet** dialog box is shown in figure *Minitab 1.1*.

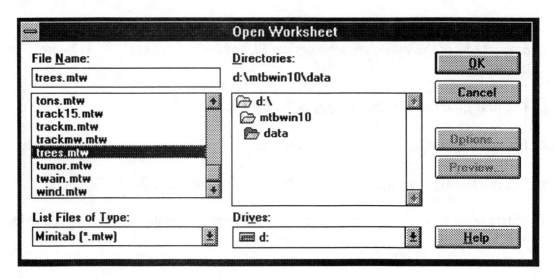

Minitab 1.1

Click on the **OK** button. The **Data** window will display the **trees** data set. Observe that the variable *HEIGHT* (of the tree) is in column C2 and there are 31 observed values.

To create the tables in **MINITAB**, select **Stat→Tables→Tally**. The **Tally** dialog box will appear. Click on C2 *(HEIGHT)* and then click on the **Select** button. The variable *HEIGHT* will appear in the **Variables** text box. Next, select **Counts**, **Percents**, **Cumulative Counts**, and **Cumulative Percents** by clicking on the boxes. An × (times symbol) in the box will indicate that you have made that selection. This is shown in figure *Minitab 1.2*. Note that **Counts** in the **Tally** box represents the frequency count and **Percents** represent the relative frequency expressed as a percent.

Note: The above procedure will only work if the values of the variable are integers.

18

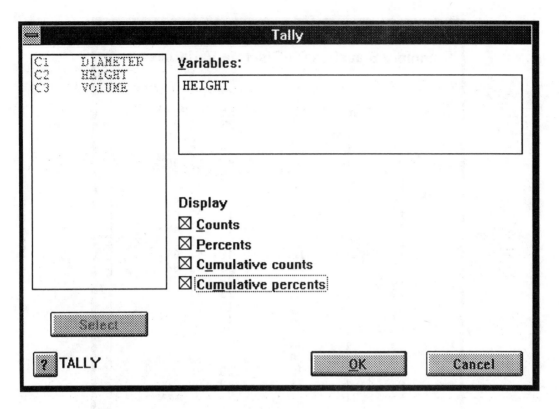

Minitab 1.2

Select the **OK** button and the tables will be displayed in the **Session** window as shown in figure *Minitab 1.3*. The first column lists the values of the variable *HEIGHT*, the second column lists the frequency counts, the third column lists the cumulative frequency counts, the fourth column lists the relative frequencies as a percent, and the last column lists the cumulative relative frequencies as a percent for the values of the variable.

```
┌─────────────────────────────────────────────────────────┐
│                        Session                          │
├─────────────────────────────────────────────────────────┤
│  Summary Statistics for Discrete Variables              │
│                                                         │
│                                                         │
│  HEIGHT  Count  CumCnt  Percent  CumPct                 │
│     63     1      1      3.23     3.23                   │
│     64     1      2      3.23     6.45                   │
│     65     1      3      3.23     9.68                   │
│     66     1      4      3.23    12.90                   │
│     69     1      5      3.23    16.13                   │
│     70     1      6      3.23    19.35                   │
│     71     1      7      3.23    22.58                   │
│     72     2      9      6.45    29.03                   │
│     74     2     11      6.45    35.48                   │
│     75     3     14      9.68    45.16                   │
│     76     2     16      6.45    51.61                   │
│     77     1     17      3.23    54.84                   │
│     78     1     18      3.23    58.06                   │
│     79     1     19      3.23    61.29                   │
│     80     5     24     16.13    77.42                   │
│     81     2     26      6.45    83.87                   │
│     82     1     27      3.23    87.10                   │
│     83     1     28      3.23    90.32                   │
│     85     1     29      3.23    93.55                   │
│     86     1     30      3.23    96.77                   │
│     87     1     31      3.23   100.00                   │
│    N=     31                                            │
└─────────────────────────────────────────────────────────┘
```

Minitab 1.3

Note: *If your computer is connected to a printer with the appropriate printer driver* **you can print any active window** *by selecting* **File→Print Window.** *Be careful to highlight the portion of the window you want to print before printing any windows. If not, the entire window will be printed and will include all unnecessary information.*

To display this set of information in the **Data** window so that you can construct appropriate graphs and charts, make the **Session** window the active window, and highlight the first row (63 1 1 3.23 1.23) to the last row (87 1 31 3.23 100.00) using the mouse. Select **Edit→Copy Cells**. Make the **Data** window the active window and select the first cell in C4. Next select **Edit→Paste/Insert Cells** and the values from *Minitab 1.3* will be displayed in columns C4 to C8.

For this presentation, C4 is renamed *HTREES* (**MINITAB** will not allow you to use the variable *HEIGHT* twice), C5 is renamed *f* for frequency, C6 is renamed *CUMf* for cumulative frequency, C7 is renamed *RELf* for relative frequency , and C8 is renamed *CUMRELf* for cumulative relative frequency. *Note:* C7 *(RELf)* and C8 *(CUMRELf)* are expressed as percents.

2. CONSTRUCTING PROJECTION (LINE) AND BAR GRAPHS, FREQUENCY POLYGONS, RELATIVE FREQUENCY GRAPHS AND PIE CHARTS

Example 1 (Continued): Use **MINITAB** to construct a projection graph for the height of the trees.

To construct a ***line*** or ***projection graph*** for the variable *HEIGHT (HTREES)*, select **Graph→Plot**. In the **Graph Variable** box select C5(f) for **Y** and C4 *(HTREES)* for **X**. In **Data Display** drop down the **Display** box and select **Project**. This sequence of commands will draw a projection graph with the frequencies along the Y-axis and the data values along the X-axis. To make it more presentable, select **Edit Attributes** and enter the appropriate **Line Type**, **Color** and **Size**. Select **Annotation→Title** to type in the title. Select **Frame→Axes** or **Frame→Tick** to darken the axes or to use an appropriate scale along the axes. A ***projection graph*** for the ***tree heights*** is shown in figure ***Minitab 1.4***.

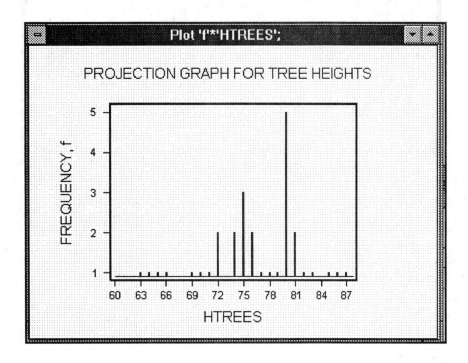

Minitab 1.4

21

Example 1 (Continued): Use **MINITAB** to construct a frequency polygon for the height of the trees.

To construct a ***frequency polygon*** for the variable *HEIGHT (HTREES)*, follow the above procedure in constructing a ***projection graph***. Select **Connect** in the **Data Display** drop down **Display** box and select **Polygon** in the **Annotation** box. The ***frequency polygon*** is shown in figure *Minitab 1.5*.

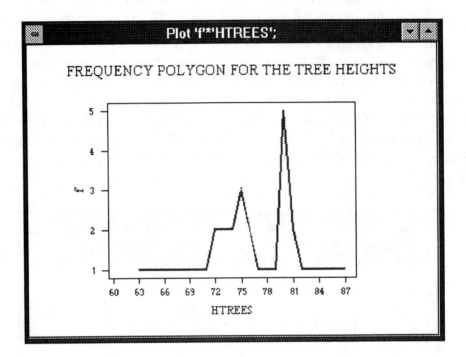

Minitab 1.5

To construct ***relative frequency*** graphs, use column C7 *(RELf; %)* instead of column C5 *(f)* in the above procedures in the construction of the projection graph and frequency polygon.

Example 2: A sample of 300 college students were asked to indicate their favorite soft drink. The survey results are shown below. Use **MINITAB** to construct a bar chart for the given information.

Soft Drink	Number of Students
Pepsi-Cola	92
Coca-Cola	78
Dr. Pepper	48
Seven-Up	42
Others	40

To construct a ***bar graph (chart)*** for the above data, enter the information in two separate columns. In this example the column for the type of soft drink was named *DRINK* and the number of students column was named *NSTUD*. Next, select **Graph→Chart** and the chart dialog box will be displayed. In the **Graph variables** box select the appropriate columns for the **X** and **Y** variables. Click on the **OK** button and the **bar graph (chart)** will be displayed as is shown in figure ***Minitab 1.6***.

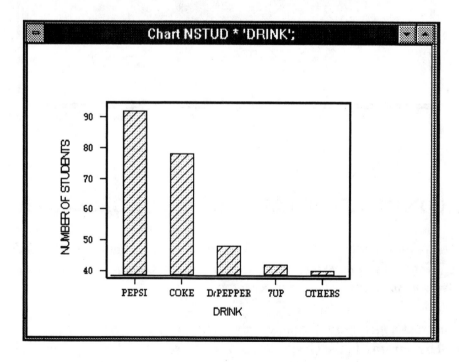

Minitab 1.6

Example 2 (Continued): Use **MINITAB** to construct a pie chart for the soft drink information.

    To construct a **pie chart** for the variable *NSTUD*, select **Graph→Pie Chart**. In the **Chart data in** box select *NSTUD*. In the **Title** box you can type in an appropriate title and in the **Options** box you can select **Add lines connecting labels to slices**. The pie chart for the tree heights is shown in figure *Minitab 1.7* with some appropriate labels.

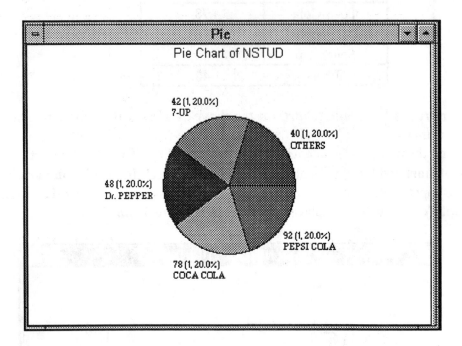

Minitab 1.7

3. CONSTRUCTING HISTOGRAMS FOR GROUPED DATA

Example 1 (Continued): Use **MINITAB** to construct a histogram for the *HEIGHT* of the trees. *Note: If you have erased the data values for* **Example 1**, *you need to open the* **trees.mtw** *worksheet again before proceeding.*

    To construct a *histogram* for the variable *HEIGHT (HTREES)*, select **Graph→Histogram**. In the **Graph variables** box, select *HTREES* and in the **Data Display** box, select **Bar**. Select **Options** and in the **Histogram Options** box, select **Frequency** for the **Type of Histogram**. For the **Type of Intervals** select **CutPoint**. This selection will place the class boundaries along the horizontal axis. Under **Definition of Intervals** you can select any of the options. For figure *Minitab 1.8* the

Midpoint/Cutpoint option was chosen with 60 : 90 / 5, i.e., the values along the X-axis ranged from 60 to 90 with class intervals of size 5. The data set was grouped first with class width of size 5.

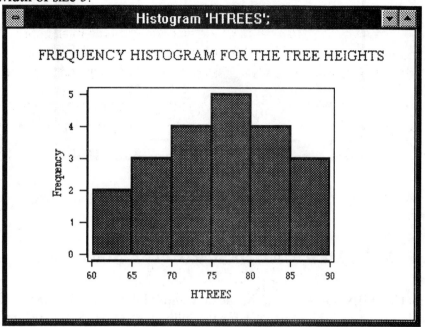

Minitab 1.8

4. CONSTRUCTING STEM-AND-LEAF PLOTS

Example 1 (Continued): Use **MINITAB** to construct a stem-and-leaf plot for the *HEIGHT* of the trees.

To construct **stem-and-leaf** plots for the variable *HEIGHT (HTREES)*, select **Graph→Character Graphs→Stem-and-Leaf**. In the **Variables** text box that appears, select *HTREES*. If you select **OK**, a stem-and-leaf display will be shown in the **Session** window. You can select different lengths of intervals for the display from the **Increment** box. For the figure *Minitab 1.9*, an increment of 5 was used.

25

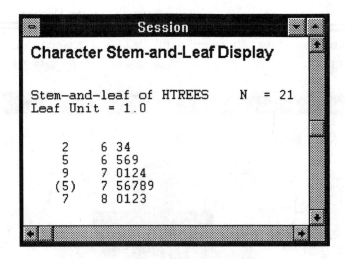

Figure 1.9

5. NUMERICAL DESCRIPTIVE MEASURES

Example 3: Use **MINITAB** to find the mean, standard deviation, variance, mode (or modes), median, first and third quartiles for the following set of 25 data values.

8.8	9.6	8.3	9.3	9.1	8.3	8.4	10.1	11.9	13.4	10.1	11.1	13
11.9	11	10	9.8	12.1	12.6	12.9	10.6	10.9	13	11.8	12.9	

    First, type the data values in column C1 in the **data** window. Select **Stat→Basic Statistics→Descriptive Statistics** and the **Descriptive Statistics** dialog box will appear. Use the mouse to highlight C1 in the left hand box by clicking on it. Next click on the **Select** button and C1 will be listed in the **Variables** text box. Alternatively, you can click on the **Variables** text box and type in C1. The **Descriptive Statistics** dialog box is shown in figure *Minitab 1.10*.

26

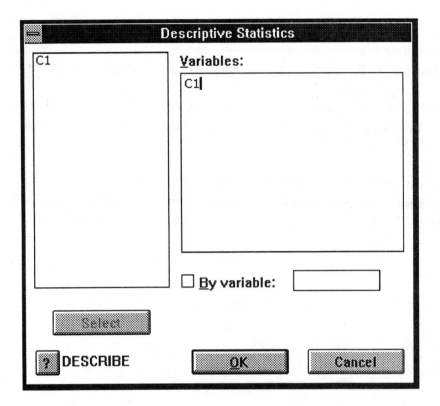

Minitab 1.10

Click on the **OK** button and the **Session** window will display the results. This window is shown in figure *Minitab 1.11*.

Descriptive Statistics

Variable	N	Mean	Median	TrMean	StDev	SEMean
C1	25	10.836	10.900	10.835	1.648	0.330

Variable	Min	Max	Q1	Q3
C1	8.300	13.400	9.450	12.350

MTB > |

Minitab 1.11

This procedure gives the values of the mean, median, standard deviation (StDev), first quartile (Q1) and the third quartile (Q3). To find the variance all you need to do is to square the standard deviation, i.e., variance $s^2 = (1.648)^2 = 2.7159$. To obtain the mode or modes, you need to construct a frequency distribution table. *The procedure use earlier*

27

will not work here because the values are not integers. However, you can construct a stem-and-leaf display to help you find the mode or modes. Verify that the modes are 8.3,

10.1, 11.9, 12.9, and 13. Note that the minimum (Min) and maximum (Max) values are also listed. You can use these values to compute Range = Maximum - Minimum = 13.4 - 8.3 = 5.1.

Note: You can compute some of these statistics individually by using **Calc→Mathematical expressions**. You can experiment with this procedure if you wish.

LAB #1: *DATA SHEET*

Name: _____   *Date:* _____

Course #: _____   *Instructor:* _____

1. **Team Exploration Project.** Use the data set named *grades.mtw* that is stored in **MINITAB** to construct:

(a) frequency tables for the variables *MATH* and *VERBAL*.
(b) projection graphs, frequency polygons, and relative frequency graphs for the variables *MATH* and *VERBAL*.
(c) histograms for the variables *MATH* and *VERBAL*.
(d) stem-and-leaf plots for the variables *MATH* and *VERBAL*.
(e) *Write up a report and discuss any observations from these graphs. Include copies of your graphs in your report.*

NOTE: Experiment with appropriate intervals when necessary to display the graphs in a clear manner.

2. **Team Exploration Project.** Collect data (from the *Wall Street Journal* etc.) for two variables that have *integer* values. Use **MINITAB** and repeat (a) through (e) as in Problem 1. Also, collect a set of qualitative data and construct a bar chart and a pie chart for the data. Present your results in a report and discuss any observations from the charts and graphs.

3. Given the following set of data use **MINITAB** to find the mean, median, standard deviation (StDev), first quartile (Q1), and the third quartile (Q3). Next, *add* 10 to each data value and repeat. *Hint*: You can let **MINITAB** help you with this mathematical operation. Suppose the data values are in C1. Select **Calc→Mathematical Expressions** and in the **Mathematical Expressions** dialog box will be displayed. In the **Variable** text box type in an appropriate column number where you want the modified values to be stored. Let this column be C2. In the Expression text box, type C2 = C1 + 10 and click on the **OK** button. A new set of values (C1 + 10) will be displayed in column C2.

60   75   81   85   90   97   103   117   125.

	Data Values	**Data Values + 10**
Mean		
Median		
Standard Deviation		
First Quartile		
Third Quartile		

Use the results from the table above, to help *generalize* the responses to the following questions.

(a) If the same constant is *added* to each value in a data set, what is the effect on the mean?

(b) If the same constant is *added* to each value in a data set, what is the effect on the median?

(c) If the same constant is *added* to each value in a data set, what is the effect on the standard deviation?

(d) If the same constant is **added** to each value in a data set, what is the effect on the first quartile?

(e) If the same constant is **added** to each value in a data set, what is the effect on the third quartile?

(f) Construct separate histograms for the original data values and the data values plus 10. Discuss your observations. Try to **generalize** as to what happens to a histogram when the same constant is added to each of the data values. *(Two sets of axes are provided for the histograms).*

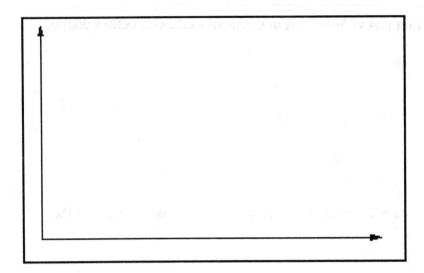

4. Given the following set of data, use **MINITAB** to find the mean, median, standard deviation (StDev), first quartile (Q1), and the third quartile (Q3). Next, *multiply* each data value by 10 and repeat. *Hint:* Follow a similar procedure as in Problem 3. In this case you can let C2 = C1*10.

50   65   71   75   80   87   93   107   115.

	Data Values	Data Values ×10
Mean		
Median		
Standard Deviation		
First Quartile		
Third Quartile		

Use the results from the above table, to *generalize* the responses to the following questions.

(a) If each value in a data set is *multiplied* by the same constant, what is the effect on the mean?

(b) If each value in a data set is *multiplied* by the same constant, what is the effect on the median?

(c) If each value in a data set is *multiplied* by the same constant , what is the effect on the standard deviation?

(d) If each value in a data set is *multiplied* by the same constant, what is the effect on the first quartile?

(e) If each value in a data set is ***multiplied*** by the same constant, what is the effect on the third quartile?

(f) Construct separate histograms for the original data values and the data values multiplied by 10. Discuss your observations. Try to ***generalize*** as to what happens to a histogram when each value in a data set is multiplied by the same constant.

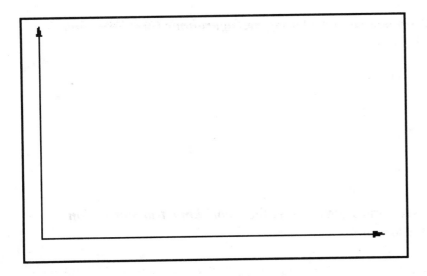

5. Given the following set of data, use **MINITAB** to find the mean, median, standard deviation (StDev), first quartile (Q1), and the third quartile (Q3).

69 56 99 125 50 65 71 113 85 75 80 87 93 107 115.

In scientific studies, sometimes it is convenient to transform the data values. One such transformation is the *logarithmic* transformation. Transform the values using a logarithmic transformation before answering the following questions.

Hint: To use **MINITAB** to transform the data before using the **Descriptive Statistics** procedure, select **Calc→Mathematical Expressions**. If the data values are stored in column C1, save the new variable in C2 and in the **Expression** text box type **Loge(C1)** or **Logten(C1)** depending on which base you want to use. When you click on the **OK** button the transformed data values will be stored in column C2.

	Data Values	LOG(Data Values)
Mean		
Median		
Standard Deviation		
First Quartile		
Third Quartile		

Use the results from the table above, to help *generalize* the responses to the following questions.

(a) If each value in a data set is transformed using the *logarithmic transformation*, what is the effect on the mean?

(b) If each value in a data set is transformed using the *logarithmic transformation*, what is the effect on the median?

(c) If each value in a data set is transformed using the *logarithmic transformation*, what is the effect on the standard deviation?

(d) If each value in a data set is transformed using the *logarithmic transformation*, what is the effect on the first quartile?

(e) If each value in a data set is transformed using the *logarithmic transformation*, what is the effect on the third quartile?

6. Given the following set of data, use **MINITAB** to find the mean, median, standard deviation (StDev), first quartile (Q1), and the third quartile (Q3). Next, change the last data value from 115 to 115,789, i.e., 115,787 is acting as an *outlier* (very large or very small value relative to the rest of the data set).

    59   91   79   63   77   71   84   87   96   103   115 (115,789).

	Data Values	**Data Values with Outlier**
Mean		
Median		
Standard Deviation		
First Quartile		
Third Quartile		

Use the results from the table above to *generalize* the responses to the following questions.

(a) If there is an *outlier* in the data set, what is the effect on the mean?

37

(b) If there is an *outlier* in the data set, what is the effect on the median?

(c) If there is an *outlier* in the data set, what is the effect on the standard deviation?

(d) If there is an *outlier* in the data set, what is the effect on the first quartile?

(e) If there is an *outlier* in the data set, what is the effect on the third quartile?

(f) Construct separate histograms for the original data values and the data values with the outlier. Discuss your observations. *Generalize* as to what happens to a histogram when an outlier is present in the data set.

39

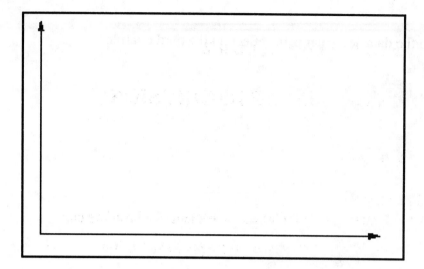

STATISTICS LAB # 2

BIVARIATE DATA - LINEAR REGRESSION

PURPOSE - to use MINITAB to

1. display a **scatter plot** for bivariate data
2. determine and interpret the **Pearson r correlation coefficient** for bivariate data
3. produce the equation $\hat{y} = mx + b$ of the **least-squares (regression) line** for bivariate data
4. graph and analyze the least-squares line on a scatter plot
5. **estimate** $\hat{y}$ for a given value x of the independent variable

BACKGROUND INFORMATION

1. **Bivariate data** - data pairs (x, y) which have been collected for statistical analysis. Call x the **independent variable** and y the **dependent variable**.

2. **Scatter plot** - a plot of the bivariate data pairs in the plane.

3. **Linear correlation of x with y -**
   (a) positive linear correlation: y tends to increase linearly as x increases;
   (b) negative linear correlation: y tends to decrease linearly as x increases.

4. **Pearson r correlation coefficient** - a value between -1 and +1 which tests the degree of linear correlation of x with y. The value of r is found from the equation

$$r = \frac{SS_{xy}}{\sqrt{SS_x SS_y}}$$

where

Sum of squares for x, $SS_x = \dfrac{\Sigma(x - \bar{x})^2}{n} = \Sigma x^2 - \left(\Sigma x\right)^2 / n,$

Sum of squares for y, $SS_y = \dfrac{\Sigma(y - \bar{y})^2}{n} = \Sigma y^2 - \left(\Sigma y\right)^2 / n,$

41

and

Sum of squares for xy, $SS_{xy} = \dfrac{\Sigma(x-\bar{x})(y-\bar{y})}{n} = \Sigma xy - \left(\Sigma x\right)\left(\Sigma y\right)/n.$

Interpretation of r: When r is

  (a) close to +1 $\Rightarrow$ a strong positive linear correlation of x with y;
  (b) close to -1 $\Rightarrow$ a strong negative linear correlation of x with y;
  (c) close to 0 $\Rightarrow$ a weak linear correlation of x with y.

5. **Linear Regression** - fitting the bivariate data with the least-squares line $\hat{y} = mx + b.$

This line has the smallest sum of the squares of errors $SSE = \Sigma(y-\hat{y})^2$, where the sum is over the (x, y) data pairs.

6. A **residual** - The difference $y - \hat{y}$ for a given data pair (x, y).

7. The **slope m and y-intercept b of the least-squares line** - they are given by $m = SSxy / SSx$ and $b = \bar{y} - m\bar{x}.$ The point $(\bar{x}, \bar{y})$ is always located on the least-squares line.

8. **Using $\hat{y}$ as an estimator of y** - the value of $\hat{y}$ obtained from the least-squares (regression) line $\hat{y} = mx + b,$ for a given value of x.

PROCEDURES

Bivariate data consists of a collection of data pairs (x, y), where x is the independent variable and y is the dependent variable. To work with such pairs, enter **MINITAB** by following the instructions provided in *Lab #0*. Enter the x-values in one column and the y-values in a second column in the **Data** window. This data can then be analyzed using a scatter plot, generating various statistical measures, and testing for a linear correlation of x with y. The regression line can then be used to generate an estimate $\hat{y}$ of y for a given value of x.

1. ENTERING BIVARIATE DATA AND COMPUTING BIVARIATE STATISTICS

Example 1: Ten adult males were polled to determine the relationship between height x (in inches) and weight y (in pounds).

Height	70	66	68	74	69
Weight	190	155	195	230	200

Height	65	70	67	69	75
Weight	148	166	152	192	210

This data can be entered in **MINITAB** by first typing the x-values in column C1 (label as *Height*) and corresponding y-values in column C2 (label as *Weight*).

MINITAB can display statistical information about each variable such as the mean, standard deviation, median, and quartiles. Click on the menu commands and select **Stat→Basic Statistics→Descriptive Statistics** and the **Descriptive Statistics** dialog box will be displayed. Use the mouse to highlight C1 and C2 and then click on the **Select** button. The two variables, Height and Weight will be displayed in the **Variables** text box. The Descriptive Statistics dialog box with the appropriate entries is displayed below in figure ***Minitab 2.1***.

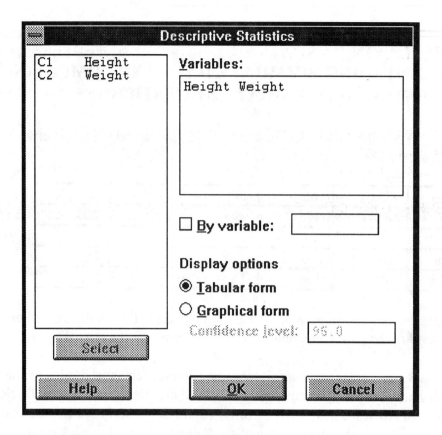

Minitab 2.1

When the **OK** button is clicked, figure *Minitab 2.2* reveals the desired statistical information which is displayed in the **Session** window.

```
┌─────────────────────────────────  Session  ──────────────────────────┐
│                                                                       │
│ Descriptive Statistics                                                │
│                                                                       │
│ Variable         N       Mean     Median    TrMean     StDev    SEMean│
│ Height          10      69.30      69.00     69.13      3.20      1.01 │
│ Weight          10     183.80     191.00    182.50     27.39      8.66 │
│                                                                       │
│ Variable        Min       Max        Q1        Q3                     │
│ Height        65.00     75.00     66.75     71.00                     │
│ Weight       148.00    230.00    154.25    202.50                     │
│                                                                       │
└───────────────────────────────────────────────────────────────────────┘
```

Minitab 2.2

Observe from the **Session** window that the mean and standard deviation of the x-values and y-values are: $\bar{x} = 69.3$ inches, $s_x = 3.2$ inches, $\bar{y} = 183.8$ pounds, $s_y = 27.39$ pounds.

44

2. DISPLAYING A SCATTER PLOT FOR BIVARIATE DATA

To demonstrate a display of a scatter plot with **MINITAB**, again refer to *Example 1*.

With the mouse select **Graph→Plot** and the **Plot** dialog box will appear. Fill in the appropriate boxes as shown in figure *Minitab 2.3*. Observe that we are letting the variable *Weight* be the dependent variable (Y) and *Height* be the independent variable (X).

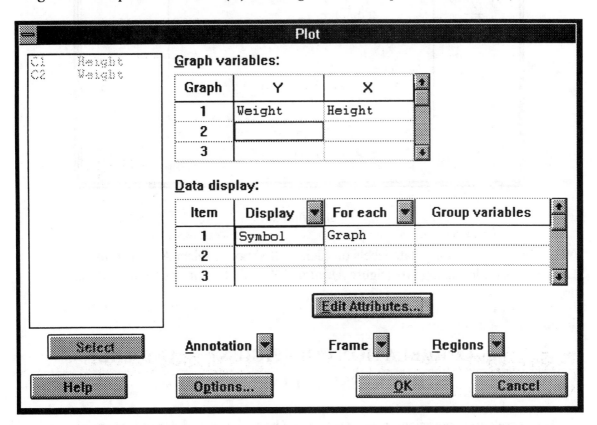

Minitab 2.3

By selecting **Edit Attributes** in the **Plot** dialog box, you can choose an appropriate symbol for your plot. The + sign was chosen in this case. Select the **OK** button and the scatter plot will be displayed in the **Plot** window as shown in figure *Minitab 2.4*.

45

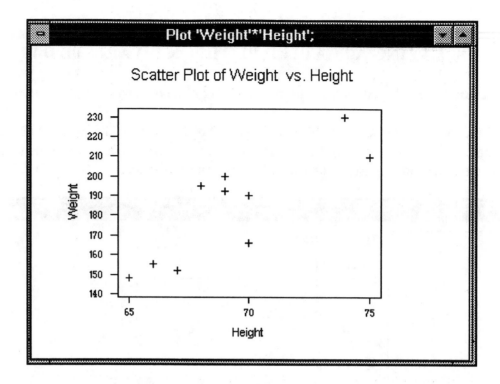

Minitab 2.4

Note: In order to display data labels on each plotted point, select **Annotation→Data Labels** in the **Plot** dialog box (figure *Minitab 2.3*) and click on the **Show data labels** box.

3. CORRELATION COEFFICIENT r AND THE
REGRESSION LINE $\hat{y}$ = mx + b

The correlation coefficient r is a measure of how closely the dependent variable y correlates with the independent variable x in a linear manner. **MINITAB** produces this value through the menu options by selecting **Stat→Basic Statistics→Correlation**. The **Correlation** dialog box with the appropriate entries is shown in figure *Minitab 2.5*.

46

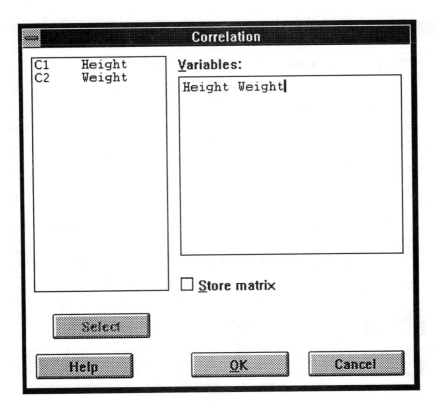

Minitab 2.5

Select the **OK** button and the value of the correlation coefficient will be given in the **Session** window. Figure *Minitab 2.6* shows a portion of this window with the computed value. The computed value is r = 0.835 which implies a relatively strong positive correlation between the variables height and weight (weight and height).

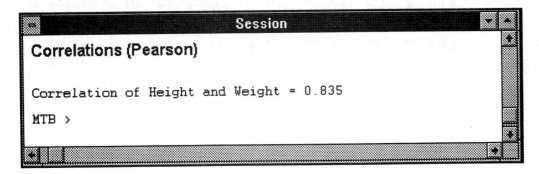

Minitab 2.6

The equation of the regression line $\hat{y}$ = mx + b is obtained by selecting **Stat→ Regression→Regression** and the **Regression** dialog box will be displayed. Figure *Minitab 2.7* shows this dialog box with the appropriate entries. The **Response** variable (Y) is *weight* and the **Predictor** variable (X) is *height*.

```
┌─────────────────────────────────────────────────────────────────┐
│ ─                          Regression                             │
├─────────────────────────────────────────────────────────────────┤
│ C1    Height      Response:      ┌────────────┐                   │
│ C2    Weight                     │ Weight     │                   │
│                                  └────────────┘                   │
│                   Predictors:  ┌─────────────────────────────┐    │
│                                │ Height │                    │    │
│                                │                             │    │
│                                │                             │    │
│                                │                             │    │
│                                │                             │    │
│                                │                             │    │
│                                └─────────────────────────────┘    │
│                                                                   │
│                   Storage                 ☐ Hi (leverage)         │
│                   ☒ Residuals             ☐ DFITS                 │
│                   ☐ Standard. resids.     ☐ Cook's distance       │
│                   ☒ Fits                  ☐ MSE                    │
│                   ☐ Coefficients          ☐ X'X inverse           │
│      [ Select ]   ☐ Deleted t resids.     ☐ R matrix              │
│                                                                   │
│  [  Help  ]    [  Options...  ]    [   OK   ]    [  Cancel  ]      │
└─────────────────────────────────────────────────────────────────┘
```

Minitab 2.7

Observe that the **Residuals** and **Fits** boxes were selected in figure *Minitab 2.7*. As a consequence, **MINITAB** has created two columns of data values in C3 (named *FITS1*) and C4 (named *RESI1*). The *FITS1* column consists of the estimates $\hat{y}$ of y and the *RESI1* column contains the residuals $y - \hat{y}$ from the computed least-squares line, using the x-values from column C1. Select the **OK** button and the regression equation will be displayed in the **Session** window. Figure *Minitab 2.8* shows a portion of the **Session** window.

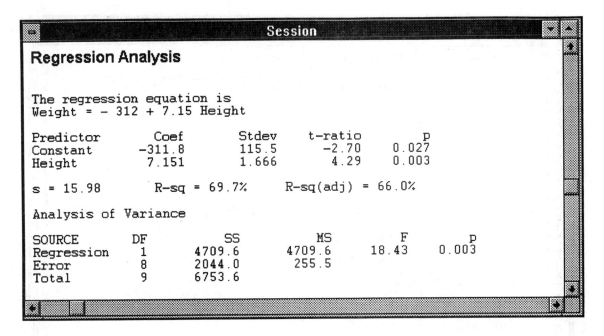

Minitab 2.8

The **Session** window exhibited in figure *Minitab 2.8* reveals that the regression line is $\hat{y} = 7.15x - 312$, with slope $m = 7.15$, and y-intercept $b = -312$.

4. GRAPHING THE REGRESSION LINE ON THE SCATTER PLOT

For a regression line graph for the height-weight data pairs, select **Stat→Regression→Fitted Line Plot** and the **Fitted Line Plot** dialog box will be displayed. In the **Response (Y)** text box select C2 and in the **Predictor (X)** text box select C1. Click on the **OK** button and the **Fit Line** window will display the fitted regression line superimposed on the scatter plot as shown in figure *Minitab 2.9*. Also, the regression equation and the square value for r are displayed.

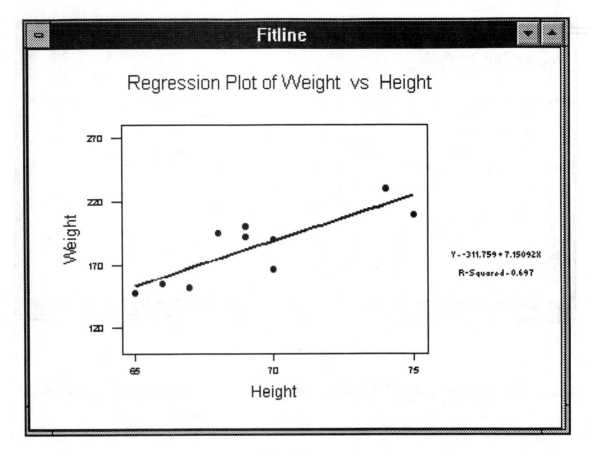

Minitab 2.9

Observe that the graph of the regression line lends support to the earlier statement of there being a positive linear correlation between height and weight.

The value of SSE = sum of the squares of the residuals (errors) can be evaluated in **MINITAB** and displayed in the **Data** window. Recall that SSE is the smallest sum among all such sums for lines used to estimate the bivariate data. The SSE value is given in figure *Minitab 2.8* as SS(Error) = 2044.

Alternatively, we can select the menu options **Calc→Mathematical Expressions**, type *SSE* in the **Variable** box, and type *sum(C4**2)* in the **Expression** box. The **Data** window now contains the information displayed in figure *Minitab 2.10* relating to the height/weight data. Observe that the value of SSE now appears in column C5.

	C1	C2	C3	C4	C5	C6	C7
↓	Height	Weight	FITS1	RESI1			
1	70	190	188.806	1.1944	2044.00		
2	66	155	160.202	-5.2020			
3	68	195	174.504	20.4962			
4	74	230	217.409	12.5907			
5	69	200	181.655	18.3453			
6	65	148	153.051	-5.0510			
7	70	166	188.806	-22.8056			
8	67	152	167.353	-15.3529			
9	69	192	181.655	10.3453			
10	75	210	224.560	-14.5603			
11							

Minitab 2.10

5. ESTIMATING y WITH THE REGRESSION LINE

Returning to *Example 1*, use the regression line $\hat{y} = 7.15x - 312$ to provide estimates of the weights (pounds) of male adults for the following heights (in inches):

64	65	66	67	68	69	70	71	72	73

Type these values in column C6 (label as *x-value*) and label column C7 as *Approx y*. Select **Calc→Mathematical Expressions** and the **Mathematical Expressions** dialog box will appear. Fill the boxes as in figure *Minitab 2.11*. Observe that the expression *7.15∗C6 - 312* from the regression line is to be typed in the **Expression** box.

Mathematical Expressions

C1 Height	**Variable (new or modified):** c7
C2 Weight	**Row number (optional):**
C3 FITS1	**Expression:**
C4 RESI1	`7.15*c6-312`
C5	
C6 x-value	

Type an expression using:

Count	Min	Absolute	Sin	+ - * / ** ( )
N	Max	Round	Cos	= <> < <= > >=
Nmiss	SSQ	Signs	Tan	AND OR NOT
Sum	Sqrt	Loge	Asin	Nscores
Mean	Sort	Logten	Acos	Parsums
Stdev	Rank	Expo	Atan	Parproducts
Median	Lag	Antilog		

Select   Help   OK   Cancel

Minitab 2.11

The **Data** window will now display the estimated (predicted) weights for these 10 heights in C7 as shown in figure *Minitab 2.12*.

Data

	C2	C3	C4	C5	C6	C7	C8
↓	Weight	FITS1	RESI1		x-value	Approx y	
1	190	188.806	1.1944	2044.00	64	145.60	
2	155	160.202	-5.2020		65	152.75	
3	195	174.504	20.4962		66	159.90	
4	230	217.409	12.5907		67	167.05	
5	200	181.655	18.3453		68	174.20	
6	148	153.051	-5.0510		69	181.35	
7	166	188.806	-22.8056		70	188.50	
8	152	167.353	-15.3529		71	195.65	
9	192	181.655	10.3453		72	202.80	
10	210	224.560	-14.5603		73	209.95	
11							

Minitab 2.12

52

LAB #2: *DATA SHEET*

Name: _____   *Date:* _____

Course #: _____   *Instructor:* _____

1. Consider the bivariate data as described in the following table:

x	1	1	3	4	4	6	7	9
y	8	6	3	3	2	2	0	1

Answer the following without the aid of **MINITAB**. This will familiarize you with the traditional procedure for finding the correlation coefficient and regression line.

(a) Construct a scatter plot for this data. Provide a scale on the x-axis and y-axis.

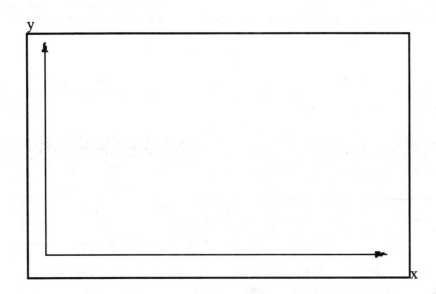

(b) Calculate the means $\bar{x} = (\Sigma x)/n$ and $\bar{y} = (\Sigma y)/n$ of the x-values and y-values. Plot $(\bar{x}, \bar{y})$ on your scatter plot in part (a).

(c) Recall that the values of the correlation coefficient r, the slope m, and y-intercept b for the regression line are provided as **BACKGROUND INFORMATION** at the beginning of this chapter. Calculate the following:

$SS_x =$ _____ $SS_y =$ _____ $SS_{xy} =$ _____

r = _____ m = _____ b = _____

2. Consider the same bivariate data as given in Problem 1:

x	1	1	3	4	4	6	7	9
y	8	6	3	3	2	2	0	1

Work this problem by using **MINITAB**. Explain your procedure for each part. Compare your answers with those found in Problem 1.

(a) Enter the x-values and y-values in columns C1 (label: x) and C2 (label: y). Apply **Stat→Basic Statistics→Descriptive Statistics** to determine the means and standard deviations of the x-values and y-values. Display the **Session** window as described in figure *Minitab 2.2*.

$\bar{x} =$ _____ $s_x =$ _____ $\bar{y} =$ _____ $s_y =$ _____

(b) Find the values of r, m, and b with the menu commands **Stat→Basic Statistics →Correlation** and **Stat→Regression→Regression.** Provide a display of the **Session** window as described in figure *Minitab 2.7.* A column of $\hat{y}$ values should now be found in C4 (as *FITS1*) and a the residuals y - $\hat{y}$ should be in C5 (as *RESI1*).

r = _____ . m = _____ . b = _____ . Equation: $\hat{y}$ = _____ .

Indicate the type of linear correlation of y with x:

(c) Apply **Graph→Plot** to display a scatter plot and regression line as described in figure *Minitab 2.8.* Based on the relationship of this line with the scatter plot, how would you describe the nature of the linear correlation of y with x?

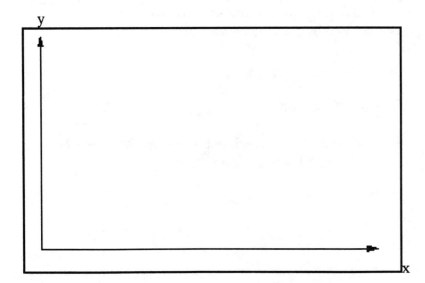

(d) Select **Calc→Mathematical Expressions** to generate the sum of squares of the residuals in column C5. Then enter the integers 1 through 10 in column C6 (label: *x-value*) and label C7 as *Approx y*. See figure **Minitab 2.11** for the procedure to generate $\hat{y}$ values in column C7. Provide a display of the **Session** window as described in figure **Minitab 2.12** by highlighting the appropriate cells and print to a printer. Remember to enter the expression from the regression line in the **Expression** box.

3.   The following bivariate data was gathered relating to the high temperature reached on a given day and the quantity of cans of soft drinks sold from a vending machine. Data was collected over a three week period.

Temperature	70	75	80	90	93	98	72	75
Quantity	30	28	40	52	57	54	27	38

Temperature	75	80	90	95	98	91	98	
Quantity	32	46	49	51	62	48	58	

Use the various **MINITAB** menu options discussed in *Example 1* to work the following parts of this exercise.

(a) What was the average daily high temperature and average number of cans of soft drinks dispensed by the vending machine?

Average daily high temperature: _____.

Average number of cans of soft drink: _____.

(b) Based on the value of the correlation coefficient r, explain how daily high temperature is linearly correlated with soft drink sales.

Value of r: _____.

Explanation:

(c) Find the slope m and the y-intercept b of the regression equation. Write down the equation of this line.

m = _____.   b = _____. $\hat{y}$ = _____.

(d) Use **MINITAB** to provide a printout of a **Graph** window of a scatter plot containing a graph of the regression line. Based on this graph, how would you describe the nature of the linear correlation between temperature and sales?

(e) Estimate the number of soft drinks dispensed for the following table of daily high temperatures. Use the methods as described in figure *Minitab 2.11* and figure *Minitab 2.12*.

Temperature	74	76	78	80	82	84	86	88	90	92
Pop Sales										

57

4. **Team Exploration Project** - Design an experiment to select random samples of 50 males and 50 females and compare height (x) with shoe size (y).

(a) Explain the procedure used to gather the data. Display these values.

Use **MINITAB** in producing the following information.

(b) Determine the mean and standard deviation of the heights and of the shoe sizes for both the male sample and female sample.
(c) Determine the correlation coefficient r, slope m, y-intercept b, and equation of the regression line for both samples.
(d) Display a scatter plot of the bivariate data, together with a graph of the regression line for both samples.
(e) Estimate the shoe sizes of a male and female for a range of 15 heights of your choice, where the difference between two consecutive heights is one inch.
(f) Construct histograms for the heights and shoe sizes for both samples and comment on their shapes.

Finally, write a report which summarizes what you have discovered about the relationship between height and shoe size, for males and females, based on the information produced from your study using **MINITAB**. Compare your results for the male and female groups.

STATISTICS LAB # 3

PROBABILITY - SIMULATION OF EXPERIMENTS

PURPOSE - to use MINITAB to

1. generate **random numbers** from a variety of population distributions
2. perform simulations of **probability experiments**
3. aid in the understanding of the **Law of Large Numbers**
4. display graphs relating to **theoretical** and **empirical probability** distributions
5. develop an understanding of an experiment involving **independent trials**

BACKGROUND INFORMATION

1. **Probability experiment** - a **random process** which generates various results called **outcomes** of the experiment. In a random process the results will not be known in advance but should follow a predictable pattern when the experiment is repeated a large number of times.

2. **Sample space** - the set of all outcomes of a probability experiment.

3. **Event** - a subset of the sample space.

4. **Probability of an event E** - a value P(E) between 0 and 1 which is assigned to E. If the outcomes in E are *equally likely* to occur, then
   $$P(E) = \frac{f}{n} \quad \text{where}$$
   n = number of outcomes in the sample space and
   f = number of outcomes belonging to E.

5. **Empirical Probability of an event E** - if a probability experiment is carried out repeatedly, then the empirical probability of E is given by
   $$P'(E) = \frac{f}{n} \quad \text{where}$$
   n = number of trials of the experiment,
   f = number of "favorable" trials, i.e., trials where the outcome is in E.

6. **Law of Large Numbers** - when a probability experiment is conducted a large number of times, the empirical probability P'(E) can be expected to be close to the theoretical probability P(E) of an event E. This approximation to P(E) should improve as the number of trials increases.

7. **Independent trials** - when a probability experiment involves several trials, they are independent whenever a probability associated with one trial is not affected by what has occurred from the other trials.

8. **Random number** - a number chosen in a random manner from a given probability distribution. When a large number of random selections are made, the distribution of the resulting values should closely conform to given theoretical probability distribution.

9. **Simulation of a probability experiment** - a model used to generate random numbers which represent the outcomes of a probability experiment.

10. **Bernoulli population** - a population consisting of two types of outcomes, called "successes" and "failures". A random selection from this population will result in a success with probability p and a failure with probability 1 - p.

PROCEDURES

When the **MINITAB** option **Calc→Random Data** is chosen, figure *Minitab 3.1* shows the various distributions from which to select to conduct random sampling.

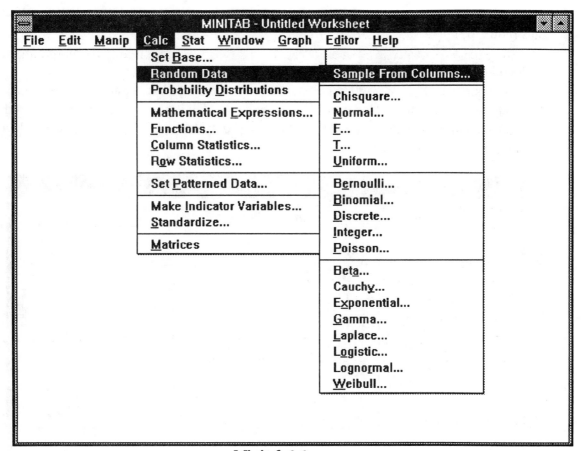

Minitab 3.1

Several of these distributions will be utilized in the simulation of probability experiments, such as coin and dice tossing. Of particular interest for the simulation of probability experiments are the **Sample from Columns**→**Bernoulli**, **Discrete**, and **Integer** options.

1. RANDOM NUMBERS GENERATED FROM A BERNOULLI POPULATION

A Bernoulli population consists of "successes" and "failures. The numbers 1 and 0 will represent a "success" and "failure" being selected, respectively. If E is the outcome where a 1 occurs (for "success"), then when there are n trials and f 1-values, the empirical probability of success is $P'(E) = f / n$ which also represents the average number of successes in the n trials. The following example demonstrates how **MINITAB** generates these random numbers.

Example 1: Simulate the tossing of a fair coin 500 times and record heads (code: 1) and tails (code: 0).

Label column C1 with *Toss No.* and column C2 with *Value*.
Enter the numbers 1, 2, ..., 500 in column C1 by selecting **Calc→Set Patterned Data**. The **Set Patterned data** dialog box will be displayed. Fill in the dialog box as indicated in figure *Minitab 3.2*. Click on the **OK** button and the values 1, 2, ..., 500 will be displayed in column C1.

Minitab 3.2

Next, to generate 500 values from the Bernoulli distribution into column C2, select **Calc→Random Data→Bernoulli**. In the **Bernoulli Distribution** dialog box that appears, enter the information as shown in figure *Minitab 3.3*.

```
┌─────────────────────────────────────────────────────────────────┐
│ ▓▓                      Bernoulli Distribution                    │
├─────────────────────────────────────────────────────────────────┤
│ C1    Toss No.    Generate  ┌─────┐     rows of data              │
│ C2    Value                 │ 500 │                               │
│                             └─────┘                               │
│                   Store in column(s):                            │
│                   ┌───────────────────────────────────────┐      │
│                   │ C2 │                                   │      │
│                   │                                        │      │
│                   │                                        │      │
│                   │                                        │      │
│                   └───────────────────────────────────────┘      │
│                                                                   │
│                   Probability of success:     ┌─────────┐         │
│                                               │  0.5    │         │
│                                               └─────────┘         │
│                                                                   │
│         ┌──────────┐                                              │
│         │  Select  │                                              │
│         └──────────┘                                              │
│  ┌──────────┐                    ┌──────────┐   ┌──────────┐      │
│  │   Help   │                    │    OK    │   │  Cancel  │      │
│  └──────────┘                    └──────────┘   └──────────┘      │
└─────────────────────────────────────────────────────────────────┘
```

Minitab 3.3

The effect of the above commands is to randomly generate 500 0-values and 1-values in column C2. Note that in the **Probability of Success** text box, a value of 0.5 is used. Thus we are simulating the tossing of a fair coin 500 times.

The **Law of Large Numbers** will now be explored for this coin toss experiment. Recall that this means that as the number of trials of a probability experiment increases, the empirical probability $P'(E)$ for a given event E should approach the theoretical probability $P(E)$.

In this exercise the event E represents a head (1-value) occurring. The value of $P'(E)$ should approach the theoretical probability $P(E) = 0.5$ for large values of n. In order to display the values of $P'(E)$ for $n = 1, 2, ..., 500$, carry out the following **MINITAB** steps.

Place a label of *Par Sums* for column C3 and *Av Heads* for column C4. Select **Calc→Functions** and fill in the values in the **Functions** dialog box as shown in figure *Minitab 3.4*. Note that the **Partial Sums** button was selected.

Minitab 3.4

As a consequence of the above procedure, column C3 will now contain a running total of the number of heads appearing in the simulation for n trials where n = 1, 2, 3, ... ,500.

Next, a column of empirical probabilities P'(E) of a head are generated in column C4 through the **MINITAB** commands **Calc→ Mathematical Expressions** (see figure *Minitab 2.11*). In the **Variable** box enter *C4* and in the **Expression** box type *C3/C1*. Then C4 will contain the values P'(E) = f/n where f = number of heads in the first n trials of the experiment for n = 1, 2, 3, ... ,500.

To graph the running averages of heads for n = 1, 2, 3, ... ,500 tosses of the coin, select **Graph→Plot** and fill in the dialog box as shown in figure *Minitab 3.5*. After selecting **Annotation→Title**, type the title *Running Average of Heads*.

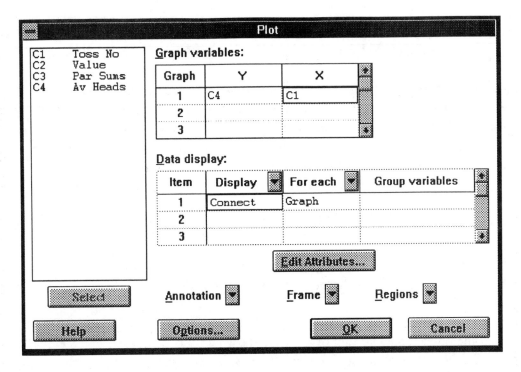

Minitab 3.5

Figure *Minitab 3.6* displays a graph of the running average of heads, i.e., empirical probabilities, for one simulation of tossing a fair coin 500 times. As stated in the **Law of Large Numbers**, as the number of tosses increases, the empirical probabilities $P'(E)$ appears to cluster around $P(E) = 0.5$ for a fair coin. For this simulation, the mean number of heads after 500 trials is approximately 0.51.

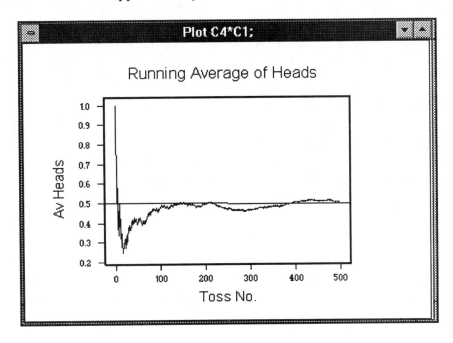

Minitab 3.6: Coin Toss Experiment

65

2. RANDOM NUMBERS FROM AN INTEGER DISTRIBUTION

The **integer distribution** assigns the same probability to any one of a sequence of consecutive integers. For example, the integer probability distribution for the consecutive integers from 3 to 7 will provide equal probabilities of 0.2 to each of the integer values from 3 to 7, as shown in the following table. The integer distribution is sometimes referred to as the **discrete uniform distribution**.

x-value	3	4	5	6	7
P(x)	0.2	0.2	0.2	0.2	0.2

A random selection from an integer distribution is useful in simulating experiments involving a sample space consisting of equally likely outcomes.

Example 2: Consider the experiment of tossing a pair of fair six-sided dice, one being white and the other red. The sum of the dots showing up is then recorded. The sample space S can be described as the 36 ordered pairs (x, y) where x and y are the number of dots up on the white and red die, respectively. The probability distribution for the sums is summarized in the following table, resulting from the 36 equally likely outcomes, given by

$$S = \{(1,1), (1,2), ..., (1,6), (2,1), (2,2), ..., (2,6), ..., (6,1), (6,2), ..., (6,6)\}.$$

By adding the two numbers in each ordered pair, the following probability distribution is obtained for the sum. Verify by listing all the points in the sample space.

Sum	2	3	4	5	6	7	8	9	10	11	12
Probability	1/36	2/36	3/36	4/36	5/36	6/36	5/36	4/36	3/36	2/36	1/36

This dice toss procedure can be viewed as an experiment consisting of two independent trials, namely tossing the white die followed by tossing the red die. **MINITAB** will be utilized in simulating this experiment 500 times.

Before proceeding, erase all data values entered in the cells from ***Exercise 1***. You can achieve this by selecting **Manip→Erase Variables** and select the appropriate variables (columns) to be deleted. Now, relabel column C1 as *White*, column C2 as *Red*, and column C3 as *Dice Sum*. To generate 500 random integers from 1 through 6, select **Calc→Random Data→Integer**, and the **Integer Distribution** dialog box will appear. Figure ***Minitab 3.7*** shows this dialog box with the appropriate entries. For each of the 500 rows, the two values in columns C1 and C2 represent the toss results for the white and red die, respectively.

Integer Distribution

Generate [500] rows of data

Store in column(s):

[c1 c2]

Minimum value: [1]

Maximum value: [6]

Select

Help   OK   Cancel

Minitab 3.7

To compute the sum of the dots on the two faces and to save in column C3, select **Calc→Mathematical Expressions**. Enter *C3* in the **Variable** text box and *C1 + C2* in the **Expression** text box. A given row entry in column C3 represents the sum of the dots showing in the toss of the white and red dice.

A display of the frequencies of the sums is now demonstrated with a histogram. First, select **Graph→Histogram** and fill in the appropriate text boxes as explained in *Lab #1*. Select **Frequency** and **MidPoint** in **Options**. Type the title *Histogram Of Dice Sum Frequencies* after selecting **Annotation→Title**. After **Annotation→Data Labels**, click on **Show data labels**. A display of the histogram is presented in figure *Minitab 3.8*. Observe that the shape of the histogram is consistent with the shape of the probability distribution from the table of probabilities presented on the previous page.

Note: If this simulation were repeated, the results are likely to be somewhat different due to randomness.

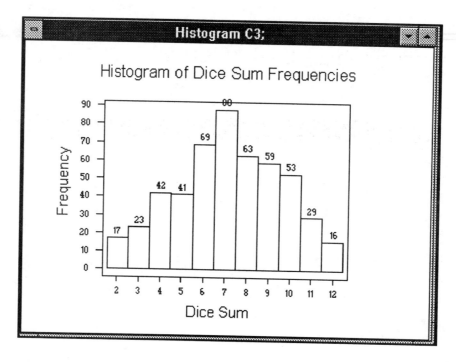

Minitab 3.8

Observe from the histogram in figure *Minitab 3.8* that a sum of four occurred 42 times compared to the expected number of occurrences: $500(3/36) \approx 42$ times.

In Problem 2 of *Lab #3*, the student will be asked to simulate this experiment and compare with the results described above.

3. PROBABILITY DISTRIBUTION GRAPHS

When a probability distribution table is given, the x-values can be stored in column C1 and their probabilities in column C2. One useful graphical display of this distribution is a **vertical projection graph**. This will be demonstrated for the dice toss problem described in *Example 2*.

Label column C4 as *Sum* and column C5 as *Prob*. Type in the sum numbers *2, 3, ..., 12* in column C4 and the corresponding probabilities *1/36, 2/36, ..., 2/36, 1/36* in column C5. *Note: These values must be entered in decimal form*. This can be effectively accomplished by first entering *1, 2, 3, 4, 5, 6, 5, 4, 3, 2, 1* in column C5. Then select **Calc→Mathematical Expressions**, type *C5* in the **Variable** box, and *C5/36* in the **Expression** box.

To produce a display of this probability distribution, select **Graph→Plot** (see figure *Minitab 3.5*), fill in the **Y** box with *C5*, the **X** box with *C4*, and select **Display→ Project**. Type in a title of *Probability Distribution Projection Graph* after selecting

Annotation→Title. Observe the similarity of the histogram display of sum frequencies in figure *Minitab 3.8* with the projection display in figure *Minitab 3.9*.

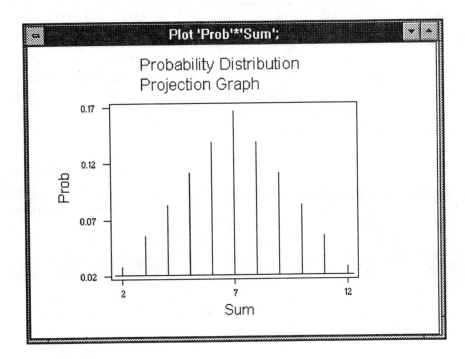

Minitab 3.9: Dice Toss Experiment

For the next exercise, a simulation is made of an experiment 1000 times. A comparison will be made of the empirical and actual probabilities.

Example 3: The charge for automotive servicing of a 4-cylinder, a 6-cylinder, and an 8-cylinder vehicle, together with percentages of these vehicle types being serviced, is presented in the following table:

No. of cylinders	4-cylinder	6-cylinder	8-cylinder
Cost of service	$32	$36	$42
Percentage serviced	25%	45%	30%

Two vehicles are randomly chosen and the average service charge recorded.

This experiment involves two independent trials, namely selecting the first vehicle and then selecting a second vehicle. The outcomes in the sample space can be viewed as pairs of vehicle servicing costs, such as (32, 36), meaning that a 4-cylinder vehicle is first chosen, followed by the selection of a 6-cylinder vehicle. From the independence of the two trials, the probability of each outcome (x, y) is the product of the probability of x with the probability of y. In the above example, $P(32, 36) = (0.25)(0.45) = 0.1125$.

69

A calculation of the outcome probabilities reveals the following probability distribution table of the average service charge for the two selected vehicles:

Average Cost	$32.00	$34.00	$36.00	$37.00	$39.00	$42.00
Probability	0.0625	0.2250	0.2025	0.1500	0.2700	0.0900

First, erase all variables in **Data** window. Next, enter the discrete probability distribution values for the service charge for a single vehicle in columns C1 and C2. Place a label of *Cost* on C1 and enter the values *32, 36,* and *42*. Label column C2 *Prob* and type in the values *0.25, 0.45,* and *0.30*.

To simulate the experiment of servicing two random vehicles and recording the average service cost, type column labels of *Cost 1* for C3, *Cost 2* for C4, and *Ave Cost* for C5. Select **Calc→Random Data→Discrete** and fill in the **Discrete Distribution** dialog box as indicated in figure *Minitab 3.10*.

Minitab 3.10

This means that for each of the 1000 times a pair of vehicles is serviced, the row entries in columns C1 and C2 represent the service costs for Vehicle 1 (C1-value) and Vehicle 2 (C2-value). For example, suppose that the values in the 50 th rows of C1 and C2 are 42 and 32, respectively. Then on the 50 th random selection of a pair of vehicles for servicing, Vehicle 1 and Vehicle 2 had service costs of $42 and $32, respectively.

To generate cost averages for the two vehicles sampled, select **Calculate→Row Statistics** and the **Row Statistics** dialog box will appear. We want to compute the average of the (pairs) of values in columns C3 and C4 and place these averages in column C5. Fill in the dialog box as shown in figure *Minitab 3.11*.

Row Statistics		

C1   Cost
C2   Prob
C3   Cost 1
C4   Cost 2
C5   Ave Cost

Statistic

○ **S**um   ○ M**e**dian
◉ **Mean**   ○ Sum of s**q**uares
○ **S**tandard deviation   ○ **N** **t**otal
○ M**i**nimum   ○ **N** nonmissing
○ Ma**x**imum   ○ N missing
○ **R**ange

Input **v**ariables:

C3 C4

[Select]

Store result in: C5 |

[Help]   [OK]   [Cancel]

Minitab 3.11

Note that each of the 1000 entries in column C5 represents the average service cost for a given pair of randomly selected vehicles.

A histogram can now be constructed for the frequencies of these averages by selecting **Graph→Histogram**. Type *Empirical Frequency Distribution* after selecting **Annotation →Title**. Click on **Show data labels** after selecting **Annotation→Data Labels**. To ensure that each average value is displayed on the horizontal axis of the histogram, click on the **Options** button and the **Histogram Options** dialog box will be displayed. Select **Frequency** for the **Type of Histogram** and the **Type of Intervals** to be **MidPoint**. Select **Midpoint/Cutpoint positions** for the **Definition of Intervals** choice and type *32:42/1* in the text box. What you are telling **MINITAB** to do is to construct a frequency histogram for values from 32 to 42 in increments of 1, and place these values at the midpoints of the boxes on the horizontal axis. *Note:* We know in this case that the average values lie between 32 and 42 inclusive so if we increment by 1, we will have 11 boxes drawn in the histogram, which can be easily displayed by **MINITAB**. However, you need to be very careful when you have a large set of data. You will need to specify an

71

appropriate number of intervals in that case if you want to use the procedure in figure *Minitab 3.12* appropriately.

Minitab 3.12

The frequency histogram of the average costs of servicing two vehicles is displayed in figure *Minitab 3.13*.

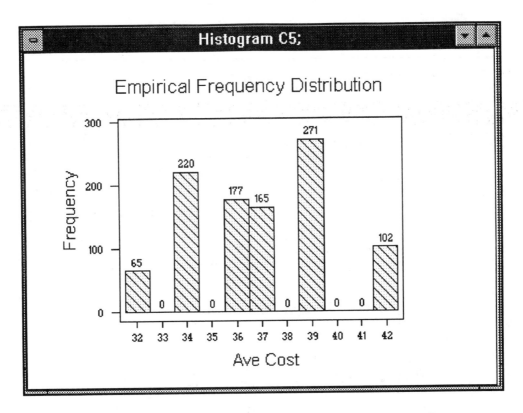

Minitab 3.13: A Simulation Of 1000 Average Service Costs

When these frequencies are converted to empirical probabilities, the following table allows for a comparison of the empirical probabilities with the true probabilities.

Average Cost	$32.00	$34.00	$36.00	$37.00	$39.00	$42.00
Empir. Prob	0.065	0.220	0.177	0.165	0.271	0.102
Probability	0.0625	0.2250	0.2025	0.1500	0.2700	0.0900

This provides another example illustrating the **Law of Large Numbers**. From the above table one observes that the empirical probability of the two vehicles having an average service cost of $34 is 0.220 compared to a theoretical probability of 0.225.

In Problem 4 of *Lab #3* the student will be asked to carry out this simulation and compare with the results described above.

NOTES

LAB #3: *DATA SHEET*

Name: _____   *Date:* _____

Course #: _____   *Instructor:* _____

1. Cathy made 68% of the free throws attempted during the previous basketball season.

   (a) By random sampling from an appropriate Bernoulli distribution, use **MINITAB** to simulate the results on her next 100 free throw attempts. Code a successful shot as a 1 and a missed shot as a 0. Generate a printout of a **Graph** window displaying a running average of shots made from 1 to 100 attempts. Explain your procedure. *Note: Due to randomness, your results on this and subsequent parts of this exercise will generally be different from others carrying out this simulation.*

   (b) Based on your display in (a), does it appear that Cathy's free throw average of successes is approaching her expected probability of success? What percentage of free throws did she make after 50 attempts?; 100 attempts?

   Percentage after 50 attempts: _____; after 100 attempts: _____.

   (c) On which toss was Cathy's free throw average the lowest?; the highest? Give the value of these percentages. Explain your answers.

   Lowest average: _____. Percentage of made shots: _____.

   Highest average: _____. Percentage of made shots: _____.

2. Use **MINITAB** to perform the simulation of tossing a white and red die as described in *Example 2*. First erase all data values from the cells in the **Data** window from Exercise 1.

(a) Type headings of *White* for column C1, *Red* for column C2, *Dice Sum* for column C3. As in *Example 2*, use **MINITAB** to generate 500 random integers from 1 through 6 in C1 and C2 and the row sums in column C3.

(b) Construct and display a histogram of the sums in C3 as described in *Example 2* (see figure *Minitab 3.8* and figure *Minitab 3.9*). Use this information to find the empirical probability $P'(E)$ that the sum of the pair of dice equals four. Compare your answer with the theoretical probability $P(E)$ that the sum equals 4. Is your answer the same as the one obtained in *Example 2*?

Empirical probability = _____.

Theoretical probability = _____.

3. Simulate the tossing of three dice 500 times using the same procedure as discussed in *Example 2*.

(a) Display a histogram of the empirical frequency distribution of the sums 3, 4, 5, ..., 18 obtained by your simulation. Discuss the shape of the histogram.

(b) From the frequencies found (a), fill in the empirical probabilities for the following table:

Sum	3	4	5	6	7	8	9	10
Empir. Prob								

Sum	11	12	13	14	15	16	17	18
Empir. Prob								

(c) The tossing of three dice consists of three independent trials. Use this information to fill in the following probability distribution table of sums:

Sum	3	4	5	6	7	8	9	10
Probability								

Sum	11	12	13	14	15	16	17	18
Probability								

As an example, observe that $P(\text{sum} = 4) = P(112 \text{ or } 121 \text{ or } 211)$
$$= 3*P(\text{first} = 1)P(\text{second} = 1)P(\text{third} = 2) = 3/216$$

(d) Compare the empirical probability $P'(E)$ that the sum is less than six with the corresponding theoretical probability $P(E)$.

$P'(E) = $ _____ . $\qquad\qquad$ $P(E) = $ _____ .

4. Simulate the vehicle servicing cost experiment as discussed in *Example 3*.

  (a) As in *Example 3*, type headings of *Cost* in column C1, *Prob* in column C2, *Cost1* in column C3, *Cost 2* in column C4, and *Ave Cost* in column C5. Then enter the probability distribution table for the service cost of one vehicle in columns C1 and C2, followed by random generation of 500 service costs for the two vehicles and their averages in columns C3, C4, and C5 (see figures *Minitab 3.10* and *Minitab 3.11*).

  (b) Construct and display a histogram of the average service costs in column C5 as described in *Example 3* (see figures *Minitab 3.12* and *Minitab 3.13*). Use your histogram to construct a table of empirical probabilities for these average service costs. Compare these average costs with the theoretical average costs.

5. A large jar is filled with nickels, dimes, and quarters. The jar contains 30% nickels, 50% dimes, and 20% quarters. An experiment consists of randomly selecting two coins and recording their total value.

  (a) Write down the set S of outcomes of this sample space as ordered pairs (x, y) where x and y are the values of the two coins selected. Find the probability of each outcome and then fill in the following probability distribution table:

  S =

Total value	$.10	$.15	$.20	$.30	$.35	$.50
Probability						

To illustrate the procedure needed to fill in the table, observe that
P(total value = $.10) = P(($0.05, $0.05)) = P(first = $0.05)P(second = $0.05)
= (0.3)(0.3) = .09.

(b) Use **MINITAB** to display a vertical projection graph of the probability distribution table from (a). Provide proper labels. Discuss the shape of the graph.

(c) Simulate this experiment 200 times by using the methods discussed in this lab session. Provide a display of a frequency histogram of total coin values for the 200 selections of a pair of coins. Provide proper labels. Discuss the shape of the histogram.

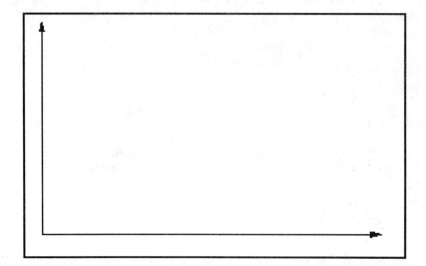

(d) From your histogram in (c), fill in the following table of empirical probabilities.

Total value	$.10	$.15	$.20	$.30	$.35	$.50
Empir. Prob.						

Compare the empirical probabilities in this table with the theoretical probabilities given in the table from part (a).

6. **Team Exploration Project - The Birthday Problem**.   In a room containing n
people, let P(n) represent the probability that two or more of them have the same
birthday. Assuming that there are 365 days in a year, the probability P(n) is described
by the following equation:

$$P(n) = 1 - \left(\frac{365}{365}\right)\left(\frac{364}{365}\right)\left(\frac{363}{365}\right)\cdots\left(\frac{366-n}{365}\right) \text{ for } n = 1, 2, ..., 366.$$

(a) Verify that this formula is correct when there are one, two, or three people in the
room.

(b) Use **MINITAB** to construct a table of these probabilities for n = 1, 2, 3, ..., 100 by
using the following procedures:
   i.   Select **Calc→Set Patterned Data** to store 1, 2, ..., 100 in column C1
         (label: *N value*)
   ii.   Select **Calc→Mathematical Expressions** to store 365/365, 364/365, ...,
         266/365 in column C2 (label : *Prob*). Enter *(366 - C1)/365* in the
         Expression: box.
   iii. Select **Calc→Function**, enter *C2* in the **Input column:**, *C2* in the **Result in:**
         box, and click on **Partial products**.
   iv.   Select **Calc→Mathematical Expressions**, enter *C2* in the **Variable:** box,
         and *1 - C2* in the **Expression:** box.

The probabilities P(n) will now be located in column C2 for n = 1, 2, ..., ,100
people.

(c) Based on the probabilities P(n) in column C2, find the smallest value n such that the
probability is 0.5 that two or more of the n people will have the same birthday. Find
the smallest n such than two or more people will have the same birthday with
probability 0.9. Explain how you arrived at your answers.

(d) Apply **Graph→Plot** to generate a printout of a graph of the probabilities for
between n = 1 and n = 100 people. Provide appropriate labels; describe what
happens as n increases.

(e) Prepare a report for this project.

NOTES

STATISTICS LAB # 4

DISCRETE PROBABILITY DISTRIBUTIONS

PURPOSE - to use MINITAB to

1. enhance understanding of **discrete random variables** and their probability distributions
2. explore through simulation the **mean (expected value)** of a discrete random variable
3. produce probabilities associated with discrete distributions, including the **binomial** and **Poisson** distributions
4. display graphs of discrete theoretical and empirical probability distributions

BACKGROUND INFORMATION

1. **Discrete random variable X** - a rule which assigns a real value x to each outcome of a probability experiment is called a **random variable**. If the x-values can be listed, X is called a **discrete random variable**.

2. **Probability distribution for a discrete random variable X** - the distribution of the x-values and associated probabilities from X.

3. **Mean (Expected value) of X** - the value $E(X) = \mu_X = \sum x\, P(x)$. The expected value of X is the mean of the population of x-values generated by X.

4. **Variance and standard deviation of X** - the variance of X is defined by
$\sigma_X^2 = \sum (x - \mu_X)^2 P(x) = \sum x^2 P(x) - \mu_X^2$; the standard deviation σ_X is the square root of the variance.

   The standard deviation and variance are measures of the dispersion of the population of x-values generated by X about the mean of X.

5. **Binomial distribution** -   the probability distribution resulting from a random variable X which counts the number of "successes" from n trials of an experiment. The trials are identical and independent and have exactly two possible outcomes, namely "success" and "failure". The probabilities of "success" and "failure" are denoted by p and q = 1 - p, respectively.

The mean, and variance of the binomial distribution are given by:

$$\mu_X = np \quad \text{and} \quad \sigma_X^2 = np(1 - p).$$

The probability distribution function is defined by:

$$P(X = x) = \binom{n}{x} p^x (1 - p)^{n-x} \quad \text{for } x = 0, 1, 2, ..., n$$

where $\binom{n}{x} = \dfrac{n!}{x!(n-x)!}$ = number of ways of selecting x objects from n objects.

6. **Poisson distribution** - a probability distribution resulting from a random variable X which counts the number of "incidents" over some increment of measurement (such as time, length, area, etc.). It is assumed that the probability of an "incident" occurring is independent of the increment chosen. The mean is denoted by λ and the probability distribution function is defined by:

$$P(x) = \lambda^x e^{-\lambda}/x! \quad \text{for } x = 0, 1, 2, 3, ...$$

PROCEDURES

The probability distribution for a discrete random variable X can be described by a listing of pairs $(x, P(x))$ where $P(x) = P(X = x)$ is the probability of X assigning x to an outcome. For any probability distribution, $\sum P(x) = 1$. Now enter **MINITAB** according to the instructions in *Lab #0*.

1. APPLYING MINITAB TO DISCRETE PROBABILITY DISTRIBUTIONS

In order to use **MINITAB** to generate the mean, standard deviation, graphs, and perform simulations on discrete random variables, it is necessary to enter the x-values in one column and the probabilities P(x) in a second column.

Example 1: Consider a discrete random variable X having the following probability distribution table:

x	1	2	3	4	5	6	7	8	9	10
P(x)	.10	.22	.28	.17	.10	.06	.04	.02	.007	.003

Recall that the mean, variance, and standard deviation of a discrete random variable X are given by:

mean of X: $\mu_X = \sum x\, P(x)$,
variance of X: $\sigma_X^2 = \sum (x - \mu_X)^2\, P(x)$,
standard deviation of X: σ_X = square root of the variance.

MINITAB can be used to find these values, as illustrated through this example. First enter the x-values in column C1, and the P(x) values in column C2. Place labels of *x-value* for column C1, *P(X=x)* for C2, *Mean* for C3, *Variance* for C4, and *Stnd Dev* for C5.

In order to find the mean μ_X, select **Calc→Mathematical Expressions** and the **Mathematical Expressions** dialog box will appear. Fill in the dialog box as shown in figure *Minitab 4.1*.

		Mathematical Expressions

C1	x-value
C2	P(X=x)
C3	Mean
C4	Variance
C5	Stnd Dev

Variable (new or modified): c3

Row number (optional):

Expression:

Sum(C1*C2)

Type an expression using:

Count	Min	Absolute	Sin	+ - * / ** ( )
N	Max	Round	Cos	= <> < <= > >=
Nmiss	SSQ	Signs	Tan	AND OR NOT
Sum	Sqrt	Loge	Asin	
Mean	Sort	Logten	Acos	Nscores
Stdev	Rank	Expo	Atan	Parsums
Median	Lag	Antilog		Parproducts

Select | Help | OK | Cancel

Minitab 4.1

Observe that *sum(C1*C2)* has been entered in the **Expression** text box. This computes the mean of X and will be displayed in the first row of column C3.

MINITAB can be used to produce the variance and standard deviation of X in the first row of columns C4 and C5, respectively. The procedure is similar to the way in which the mean of X was generated. Apply **Calc→Mathematical Expressions**, enter *C4* in the **Variable** text box, and *sum((C1-C3)**2*C2))* in the **Expression** text box. This expression represents the variance of X. Analyze this. Finally, select **Calc→Mathematical Expressions**, enter *C5* in the **Variable** text box, and *Sqrt(C4)* in the **Expression** text box.

A display of the **Data** window in figure *Minitab 4.2* reveals the values of the mean, variance, and standard deviation of X.

	C1	C2	C3	C4	C5	C6	C7
↓	x-value	P(X=x)	Mean	Variance	Stnd Dev		
1	1	0.100	3.453	3.06379	1.75037		
2	2	0.220					
3	3	0.280					
4	4	0.170					
5	5	0.100					
6	6	0.060					
7	7	0.040					
8	8	0.020					
9	9	0.007					
10	10	0.003					

Minitab 4.2

A table of cumulative probabilities $P(X \le x)$ can be generated by **MINITAB**. They are useful in finding probabilities relating to X.

Label column C6 as *P(X<=x)* and select **Calc→Probability Distributions→Discrete** and the **Discrete Distribution** dialog box will appear. Select the options as indicated in figure *Minitab 4.3*.

Minitab 4.3.

The table displayed below provides the x-values (from C1), probabilities (from C2), and cumulative probabilities (from C6).

x-value	1	2	3	4	5
P(X=x)	0.100	0.220	0.280	0.170	0.100
P(X<=x)	0.100	**0.320**	0.600	0.770	0.870

x-value	6	7	8	9	10
P(X=x)	0.060	0.040	0.020	0.007	0.003
P(X<=x)	0.930	**0.970**	0.990	0.997	1.000

From the two bold values in the above tables, the probability that X is between 3 and 7 can be computed as follows:

$$P(3 \leq X \leq 7) = P(X \leq 7) - P(X \leq 2) = 0.970 - 0.320 = 0.650$$

2. THE BINOMIAL DISTRIBUTION

The binomial random variable X has a probability distribution which can be generated by selecting **Calc→Probability Distributions→Binomial**.   The number of trials and the probability of success on a given trial must be provided.

Example 2:   A recent report concluded that 34% of adults are overweight (over the ideal weight). Suppose that a random sample of 30 adults are tested.

Let X represent the number who are overweight. Then X can be viewed as a binomial random variable with 30 independent trials. Success on a given trial means that the person tested is overweight; this probability is p = 0.34.

The probability distribution table is to be entered in columns C1 and C2, with cumulative probabilities in column C3. Provide labels of *x-value*, *P(X=x)*, and *P(X<=x)* for these columns.

Select **Calc→Set Patterned Data** (see figure *Minitab 3.2*) to enter the integers 0, 1, 2, ..., 30 in column C1. Then select **Calc→Probability Distributions→Binomial** and the **Binomial Distribution** dialog box will appear. Enter the information as displayed in figure *Minitab 4.4*.

Minitab 4.4

Again select **Calc→Probability Distributions→Binomial** and modify the input described in figure *Minitab 4.4* by clicking on **Cumulative Probability** to enter *C3* in the **Optional storage** box. A table of cumulative probabilities are now listed in column C3.

Consider the problem of finding the probability that 15 or more of these 30 people are overweight. The standard way of solving this problem is to either use a table of binomial probabilities, not applicable here since the table will not include p = .34, or by using the binomial probability formula which will require many computations.

Since column C3 contains the cumulative probability distribution with $P(X \leq 14) = 0.89661$ (in row 15), the probability that 15 or more of these 30 people are overweight is given by:

$$P(X \geq 15) = 1 - P(X \leq 14) = 1 - 0.89661 = 0.10339$$

3. GRAPHICAL DISPLAYS OF DISCRETE PROBABILITY DISTRIBUTIONS

The **Graph→Plot** option in **MINITAB** is useful in displaying several types of graphs of a discrete probability distribution. They include **Area**, **Connect**, **Project**, and **Symbol**.

Example 3: Use **MINITAB** to display a projection graph for the binomial distribution in *Example 2*. To accomplish this, select **Graph→Plot** and the Plot dialog box will appear. Select **Display→Project** in the manner explained for obtaining figure *Minitab 3.10*. The **Plot** dialog box with the appropriate entries is shown in figure *Minitab 4.5*. The title is entered after selecting **Annotation→Title**.

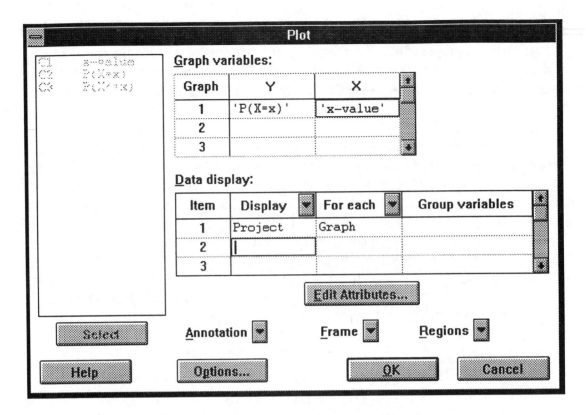

Minitab 4.5

Select the **OK** button and the graph will be displayed in the graph window. This is shown in figure *Minitab 4.6*.

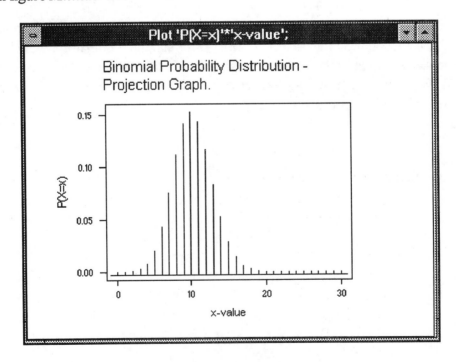

Minitab 4.6: Binomial Distribution Graph with n = 30, p = 0.34.

Observe that this probability distribution is bell-shaped and that the peak point occurs when x = 10. By applying the equation $\mu_X = np$, the mean of X is found to be 10.2 which is close to 10. The standard deviation is $\sigma_X = \sqrt{np(1-p)} \approx 2.6$. By looking at the projection probability graph in figure *Minitab 4.6*, observe that most of the distribution of x-values are within two standard deviations of the mean, represented by the interval [5, 15.4].

4. SIMULATION OF EXPERIMENTS INVOLVING DISCRETE RANDOM VARIABLES

In *Lab #3* simulations were performed of probability experiments in order to compare the theoretical probability distribution with the empirical probability distribution obtained through the simulation. In this section simulations of probability experiments will be discussed which involve various discrete probability distributions, e.g., binomial and Poisson.

MINITAB can be used to generate random data from these distributions. The process of selecting random data by applying **Calc→Random Data** has been discussed in **Lab #3**.

Example 4: A consumer agency tests new cars for defects. For a given car type, let X represent the number of defects found for a tested car. Assume that X describes a Poisson random variable and suppose that the mean number of defects is 8.5.

A comparison will be made between this probability distribution and the empirical probability distribution obtained by simulation of the experiment of inspecting 200 randomly chosen cars for the number of defects.

First, **MINITAB** will be used to calculate probabilities based on this distribution and to create a projection graph display of this distribution.

For a Poisson random variable having mean λ, the standard deviation is given by $\sigma_X = \sqrt{\lambda}$. In this example the standard deviation is approximately 2.92 so there is a high probability that an x-value is between 0 and 20.

Place headings of *x-value*, *P(X=x)*, and *P(X<=x)* for columns C1, C2, and C3. Enter the integers 0, 1, 2, ..., 20 in column C1 by selecting **Calc→Set Patterned Data**.

Select **Calc→Probability Distributions→Poisson** and the **Poisson Distribution** dialog box will appear. Fill in the information as displayed in figure *Minitab 4.7*. As a consequence of this, the probabilities P(x) for x = 0, 1, 2, ..., 20 will appear in column C2.

C1	x-value
C2	P(X=x)
C3	P(X<=x)

Poisson Distribution

◉ **P**robability
○ **C**umulative probability
○ **I**nverse cumulative probability

Mean: 8.5

◉ Input col**u**mn: c1
Optional s**t**orage: c2

○ Input co**n**stant:
Optional sto**r**age:

Select

Help   OK   Cancel

Minitab 4.7

In order to store the cumulative probabilities in column C3, apply **Calc→Probability Distributions→Poisson**, select **Cumulative probability**, and enter *C3* in the **Optional storage** box.

As in **Example 2**, a projection graph of this probability distribution is displayed in figure *Minitab 4.8* by selecting **Graph→Plot** and **Display→Project**.

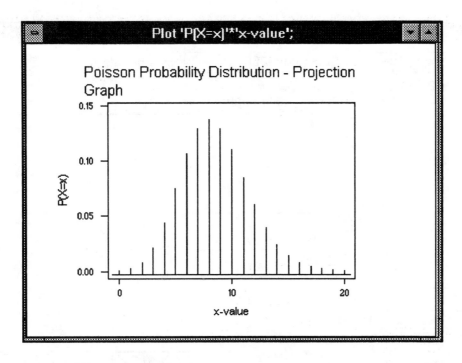

Minitab 4.8: Poisson Distribution Graph $\lambda = 8.5$

Note that this Poisson distribution is bell-shaped with mean 8.5 and a standard deviation of approximately 2.9. The peak point of the projection probability distribution graph in figure *Minitab 4.8* is at x = 8 which is close to the mean. Observe also that most of the x-values are within two standard deviations of the mean, i.e., in the interval [2.7, 14.3]. These values are consistent with the above display.

As a simulation based on *Example 4*, consider the results when the consumer agency tests the next 200 cars for defects.

Type the label *Defects* for column C4 and select **Calc→Random Data→Poisson** and the **Poisson distribution** dialog box will appear. Fill in the appropriate text boxes as indicated in figure *Minitab 4.9*.

```
┌─────────────────────────────────────────────────────────────────┐
│ ▓▓                    Poisson Distribution                       │
├─────────────────────────────────────────────────────────────────┤
│ C1    x-value      │  Generate  │ 200        │  rows of data      │
│ C2    P(X=x)       │                                              │
│ C3    P(X<=x)      │  Store in column(s):                         │
│ C4    Defects      │  ┌──────────────────────────────────────┐    │
│                    │  │ c4 |                                 │    │
│                    │  │                                      │    │
│                    │  │                                      │    │
│                    │  └──────────────────────────────────────┘    │
│                    │                                              │
│                    │  Mean:  │ 8.5        │                       │
│                    │                                              │
│   ┌──────────┐     │                                              │
│   │  Select  │     │                                              │
│   └──────────┘     │                                              │
│ ┌────────┐          ┌──────────┐      ┌──────────┐                │
│ │  Help  │          │    OK    │      │  Cancel  │                │
│ └────────┘          └──────────┘      └──────────┘                │
└─────────────────────────────────────────────────────────────────┘
```

Minitab 4.9

The command **Stat→Basic Statistics→Descriptive Statistics** when applied to the 200 defects values in column C4 results in a sample mean of 8.485 which is displayed in the **Session** window. This is close to the mean $\mu_X = 8.5$. These two values differ because we are using a random sample in one case.

Note: If this simulation were performed again, one can expect a different value for the sample mean.

A vertical projection frequency graph of the column of 200 test results is obtained by selecting **Graph→Histogram**, followed by **Display→Project** (see figure *Minitab 3.8*). Select **Options**, click on **Frequency** and **MidPoint/cutpoint position**, and type *2 : 18 / 1* in the text box. **Show data labels** has been checked following **Annotation→Data Labels** in order to list the frequency counts on the graph for the different number of defects.

A display of this frequency graph is shown in figure *Minitab 4.10*.

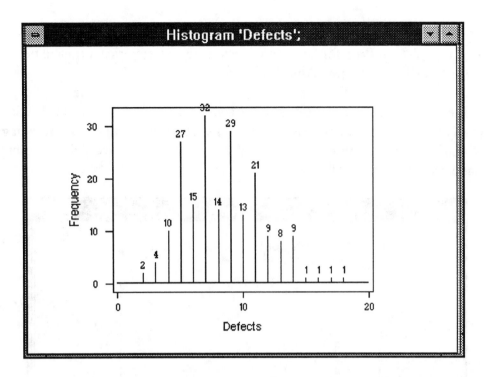

Minitab 4.10: Distribution of Defects of the 200 Cars

Compare the shape of the frequency distribution of the defects of the 200 cars in figure *Minitab 4.10* with the probability distribution display in figure *Minitab 4.8*. They are somewhat different due to the randomness of the selection procedure. If a larger number of simulations were conducted, (i.e., if a larger sample was generated), one would expect the shape of the two distributions to be in closer agreement.

A direct comparison between the empirical probability distribution of 200 test results and the probability distribution of X is found by converting the above frequencies to probabilities. For example, from figure Minitab 4.10, the empirical probability of observing 7 defects for example, is 32/200 = 0.16. This is summarized in the following table:

x-value	0	1	2	3	4	5	6
Empir. Prob	0.000	0.000	0.010	0.02	0.050	0.135	0.075

x-value	7	8	9	10	11	12	13
Empir. Prob	0.160	0.070	0.145	0.065	0.105	0.045	0.04

x-value	14	15	16	17	18	19	20
Empir. Prob	0.045	0.005	0.005	0.005	0.005	0.000	0.000

Place a label of *Emp Prob* for column C5 and enter the empirical probabilities. Select **Graph→Plot** (see figure *Minitab 3.5*).

In the **Graph** box, type *C2* and *C5* under **Y** and type *C1* and *C1* under **X**. Select **Display→Symbol** and **Frame→Multiple Graphs→Overlay graphs on same page**. In figure *Minitab 4.11*, the empirical probability distribution and the probability distribution of the random variable X are displayed.

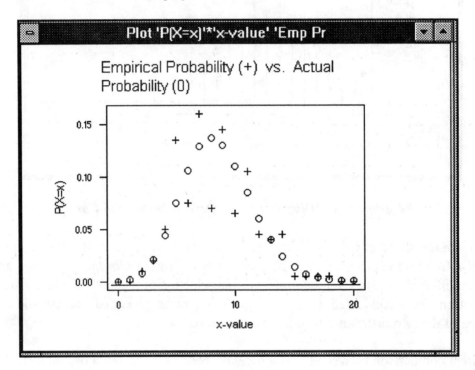

Minitab 4.11

This display demonstrates that the empirical probabilities for the number of car defects are in fairly close agreement with the theoretical probabilities. By the **Law of Large Numbers**, one could expect closer agreement with a larger number of trials (sample size).

LAB #4: DATA SHEET

Name: _____   **Date:** _____

Course #: _____   **Instructor:** _____

1. Consider a discrete random variable X which has the following probability distribution.

x value	0	1	2	3	4
P(X=x)	0.005	0.010	0.018	0.035	0.081

x value	5	6	7	8	9
P(X=x)	0.159	0.240	0.236	0.166	0.050

(a) As in *Example 1*, enter x-values in column C1, probabilities in C2, and cumulative probabilities in column C3. Use **MINITAB** to produce the mean in C4, variance in C5 and standard deviation in C6. Display the **Data** window containing these values. Provide appropriate column headings.

Mean = _____ .

Variance = _____ .

Standard deviation _____ .

(b) Use the cumulative probabilities in column C3 to find $P(3 \le X \le 8)$. Which values were used in this column in arriving at your answer?.

(c) Display a projection graph for this probability distribution, as displayed in figure *Minitab 4.5* and figure *Minitab 4.8*. Provide a hard copy of the graph. How would you describe the shape of this distribution (symmetric, skewed to the right, skewed to the left)? Also, compare the peak point on the graph with the mean and locate the two points on the x-axis which are two standard deviations below and above the mean.

(d) Perform a random selection of 1000 x-values with the commands **Calc→Random Data→Discrete** (see figure *Minitab 3.11*). Use **MINITAB** to find the mean and standard deviation of these x-values. Compare with the mean μ_X and standard deviation σ_X of X from (a).

(e) Construct and provide a printout of a frequency projection graph of these 1000 x-values using **Graph→Histogram**. (see figure *Minitab 4.10*). Explain on the basis of the randomness of the simulation how this frequency projection graph relates to the graph of the probability distribution projection graph obtained in (c).

2. Suppose that 68% of the freezers sold by an appliance store are upright models. Over the next 100 sales of freezers, let X represent the number which are uprights. Explain your procedure using **MINITAB** in arriving at your answers.

(a) Explain why X can be considered as a binomial random variable. Give the values of n = number of trials and p = probability of success on a given trial.

(b) Enter the x-values in column C1, probabilities in column C2, and cumulative probabilities in column C3. Find the probability that at least 75 of the 100 freezers sold are upright models.

(c) Calculate the mean μ_X and standard deviation σ_X of X. Determine the x-values which are within 3-standard deviation of the mean. What percentage of the total x-values does this represent?

μ_X = _____ σ_X = _____

x -values = _____   _____%

(d) Store the x-values from (c) and the corresponding probabilities in two columns. Select **Graph→Plot→Project** to display a projection graph of the probability distribution. Provide proper labels on a print out of your graph. Describe the shape of the graph and compare the x-value at the peak point with μ_X.

3. The number of calls X received by an ambulance service on a given Saturday is assumed to describe a Poisson random variable. The mean number of calls is 18.8. Explain your procedure in answering each the following.

(a) The standard deviation for a Poisson distribution is given by $\sigma_X = \sqrt{\lambda}$ where λ is the mean.   Calculate the standard deviation and determine the x-values which are within three standard deviations of the mean.

$\sigma_X =$ _____   x values: _____

(b) Generate x-values from (a) in column C1, the probabilities in column C2, and the cumulative probabilities in column C3.   What percentage of the values of X do the x-values in column C1 represent?

_____%

(c) Find the percentage of the days for which the number of calls made to the ambulance service is between 10 and 18.

(d) Construct an *area* graph of the probability distribution on column C1 and C2 by selecting **Graph→Plot** and **Display→Area**. Provide a hard copy of your graph. Determine the value of x at which the graph peaks; compare it with the mean of X.

(e) Simulate the random testing of the number of calls made over the next 100 Saturdays (see figure *Minitab 4.9*). Find the mean number of calls and compare it to μ_X.

(f) Construct a projection graph of these 100 values similar to figure *Minitab 4.10*. Select **Options**, click on **Frequency** and **MidPoint/CutPoint position**, and type *4.5 : 32.5 / 4* in the text box. Provide a hard copy of your graph. Compare the percentage of the days where the number of calls is less than 17 with the expected percentage using the cumulative probability column.

4. **Team Exploration Project.**
   (a) Generate a random sample of size 10, and using each of the probability values p = 0.05, 0.10, 0.20, 0.5, 0.7, and 0.9 for the binomial distribution. Construct projection graphs for your simulated data. Provide printouts of these graphs and discuss your observations.

   (b) Generate random samples of size n = 5, 12, 20, 30, 50, and 100 with probability p = 0.1 for a binomial distribution. Construct projection graphs for your simulated data. Provide printouts of these graphs and discuss your observations.

   (c) Generate random samples of size n = 5, 12, 20, 30, 50, and 100 with probability p = 0.5 for a binomial distribution. Construct projection graphs for your simulated data. Provide printouts of these graphs and discuss your observations.

(d) Generate random samples of size n = 5, 12, 20, 30, 50, and 100 with probability p = 0.9 for a binomial distribution. Construct projection graphs for your simulated data. Provide printouts of these graphs and discuss your observations.

Present parts (a) through (d) in a report. Formulate some general conclusions based on your observations?.

5. **Team Exploration Project.** Choose a current topic where a statistical study has been made giving a certain percentage of "successes" in a population of "successes and failures". Set up a binomial experiment with 30 trials. Put x-values in columns C1, probabilities in column C2, and cumulative probabilities in column C3. Find the probability that there are fewer than 10 successes; more than 20 successes. Display a projection probability distribution graph and locate on it the mean and points which are one standard deviation below and above the mean. Present a detailed report of your project results.

NOTES

STATISTICS LAB # 5

CONTINUOUS PROBABILITY DISTRIBUTIONS

PURPOSE - to use MINITAB to

1. enhance understanding of a **continuous random variable** and its corresponding **density function**
2. explore through simulation the **mean (expected value)**, **standard deviation**, and **shape** of various distributions of continuous random variables
3. use as an aid in producing probabilities associated with special continuous distributions such as the **Uniform**, **Normal**, and **Exponential** distributions
4. display various graphs associated with continuous random variables
5. demonstrate how well a normal distribution approximates a binomial distribution

BACKGROUND INFORMATION

1. **Continuous random variable** - a function X which assigns a real value to each outcome of a probability experiment such that the values of X range over an interval. In practice, X represents some form of measurement, such as time, length, or area.

2. **Density function for a continuous random variable X** - A function $f(x)$ defined over the real numbers having the following properties:

   (a) $f(x) \geq 0$ for all real x;
   (b) the area under the graph of $f(x)$ and above the x-axis equals one;
   (c) the probability $P(X \leq x_0)$ is the value of the area of the region in the plane which lies to the left of $x = x_0$, below the graph of f(x), and above the x-axis.

3. **Mean μ_X, variance σ_X^2, and standard deviation σ_X of X** - As with a discrete random variable, the mean is a measurement of the central tendency of x-values generated by X. The variance and standard deviations are measures of the dispersion of the distribution of x-values. Formal definitions of μ_X and σ_X^2 involve the density function $f(x)$ and the integral calculus; they will not be presented here.

4. **Uniform random variable X** - the distribution of x-values is uniform over an interval from a to b where $a < b$.

The density function $f(x)$ can be defined by

$$f(x) = \begin{cases} \dfrac{1}{b-a} & \text{if } a < x < b, \\ \\ 0 & \text{otherwise.} \end{cases}$$

The probability $P(X \le x)$ is obtained by the equation

$$P(X \le x) = \begin{cases} 0 & \text{if } x \le a, \\ \\ \dfrac{x-a}{b-a} & \text{if } a < x < b, \\ \\ 1 & \text{if } x \ge b. \end{cases}$$

The mean and standard deviation are given by $\mu = \dfrac{a+b}{2}$ and $\sigma = \dfrac{b-a}{\sqrt{12}}$, respectively.

5. **Normal random variable X** - the x-values have a bell shape distribution which is symmetric about the mean. The density function $f(x)$ is defined for real x by

$$f(x) = \frac{1}{\sqrt{2\pi}\,\sigma} e^{-((x-\mu)/\sigma)^2/2}, \quad \text{where } \mu = \text{mean and } \sigma = \text{standard deviation.}$$

The **standard normal random variable Z** has a mean of 0.0 and a standard deviation of 1.0. The transformation from X to Z is given by $Z = (X - \mu)/\sigma$.

6. **Exponential random variable X** - The distribution of x-values is skewed to the right. If $f(x)$ is the density function for X with mean μ, then

$$f(x) = \begin{cases} 0 & \text{if } x < 0, \\ \\ \left(\dfrac{1}{\mu}\right) e^{-x/\mu}, & \text{if } x \ge 0. \end{cases}$$

The standard deviation σ of X is equal to the mean μ.

PROCEDURES

The uniform, normal, and exponential distributions, as well as several others, can be addressed through **MINITAB** in order to generate random data, provide graphical displays, and evaluate probabilities.

1. RANDOM SAMPLING FROM A CONTINUOUS DISTRIBUTION

The process of random sampling from a population can provide information about the mean, standard deviation, and shape of the distribution of the data obtained. Conversely, if random sampling is performed from a population where its distribution, mean, and standard deviation are known, then the sample mean $\bar{x}$ and sample standard deviation s are expected to be in close agreement with the population mean μ and standard deviation σ, respectively. Moreover, the shape of the distribution of the data from the sample should be similar to the shape of the distribution of population values.

In this section it is demonstrated how **MINITAB** can be utilized to simulate the selection of a random sample from certain populations having continuous distributions. Sampling from discrete distributions was previously discussed in *Lab #4*. Comparisons will be made of the sampling population with the random sample of data values. First, enter **MINITAB** as explained in *Lab #0*. The **MINITAB** menu option **Calc→Random Data** enables one to randomly select data from a variety of both discrete and continuous distribution, as is demonstrated in figure *Minitab 5.1*.

Minitab 5.1

Example 1: Suppose that the amount of time that a secretary shows up late for work in the morning is approximately a uniform random variable X over the interval from -15 to 10 minutes. For example, the secretary arrives 15 minutes early when x = -15 and arrives 10 minutes late when x = 10.

The formulas for the mean and standard deviation of a uniform random variable result in the following regarding X:

$$\mu = (-15 + 10)/2 = -2.5 \text{ minutes and } \sigma = (10 + 15)/\sqrt{12} \approx 7.22 \text{ minutes.}$$

MINITAB will now be used to simulate the secretary's arrival times to work over the next 150 work days. The sample mean $\bar{x}$ and sample standard deviation s will then be generated and compared with μ and σ, respectively. Also, the shape of a histogram of these 150 arrival times will be compared to the uniform distribution of all arrival times.

Place a heading of *Sample* on column C1 and select the menu options **Calc→Random Data→Uniform**. Then enter the information called for in the **Uniform Distribution** window as described in figure *Minitab 5.2*.

Minitab 5.2

After applying **Stat→Basic Statistics→Descriptive Statistics** to the simulation of the 150 arrival times in column C1, it was found that $\bar{x} = -2.343$ and $s = 7.162$. The secretary will have arrived at work an average of 2.343 minutes early over the next 150 work days. Recall that $\mu = -2.5$, which is in close agreement with the value of $\bar{x}$. Moreover, the value $s = 7.162$ is close to the population standard deviation $\sigma = 7.22$.

Note: One should expect different results when repeating this simulation!

A histogram of these arrival times is accomplished through **Graph→Histogram**. In the **Histogram** dialog box **Options** and **Cutpoint** were selected and the interval values chosen were -15 : 10 / 5. *Interpretation:* start at -15, end at 10, and increment by 5. The histogram of this simulation is displayed in figure *Minitab 5.3*.

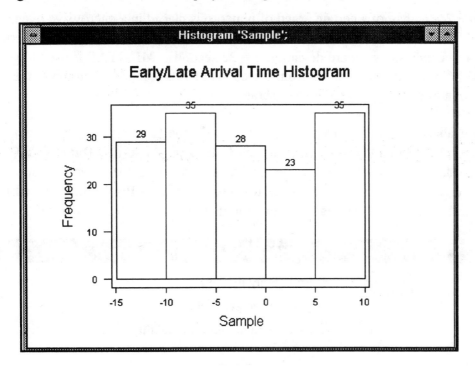

Minitab 5.3

Observe that the shape of the histogram is consistent with that of a uniform distribution of arrival times. By adding frequencies from the first three bars in the above histogram, it is found that for 92 out of the 150 days (61%), the secretary arrived to work early. The probability of early arrival based on the uniform distribution of arrival times is found by calculating the area to the left of $X < 0$: $P(X < 0) = (0 + 15)/25 = 0.60$

This says that the secretary can be expected to be early 60% of the time. The above empirical probability of 61% is in close agreement with the expected results for this uniform random variable. This again demonstrates the **Law of Large Numbers**, namely, as the number of trials (sample size) increases, the empirical probability is expected to be close to the actual probability.

2. DISPLAYS OF CONTINUOUS PROBABILITY DISTRIBUTIONS

The **MINITAB** options **Calc→Probability Distributions** enable one to generate a table of density function values for many of the common continuous distributions when the **Probability Density** option is selected. From the columns of density values, the **MINITAB** commands **Graph→Plot** can be used to display a graph of the given continuous distribution.

Example 2: A bag of a certain brand of popcorn is placed in a microwave oven. Suppose that the time X it takes for a kernel to pop has a normal distribution with mean $\mu = 114$ seconds and standard deviation $\sigma = 22$ seconds. **MINITAB** is used to generate a table of density function values from 0 seconds to 180 seconds. A graph of the distribution of popping times is also generated.

Place labels of *Time* and *Density* on columns C1 and C2, respectively. A column of popping times in column C1 is obtained from **Calc→Set Patterned Data** with 0 as the starting value, 180 as the ending value, and an increment of 1.0. The corresponding density function values in C2 are generated through **Calc→Probability Distributions →Normal** by filling the **Normal Distribution** dialog shown in figure *Minitab 5.4*.

Normal Distribution dialog box:

- C1 Time
- C2 Density

- ● Probability density
- ○ Cumulative probability
- ○ Inverse cumulative probability

Mean: 114
Standard deviation: 22

- ● Input column: C1
  Optional storage: C2
- ○ Input constant:
  Optional storage:

Select | ? PDF | OK | Cancel

Minitab 5.4

A display of the distribution of popping times shown in figure *Minitab 5.5* is accomplished through **Graph→Plot**, followed by **Display→Connect**.

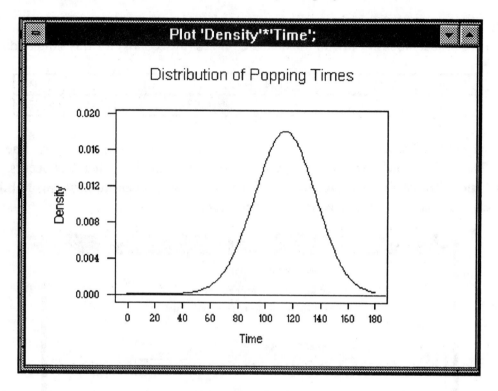

Minitab 5.5

Note that the greatest popping frequency occurs around the mean of 114 seconds and that little popping activity occurs over the first 60 seconds and after 160 seconds.

3. FINDING PROBABILITIES AND INVERSE PROBABILITIES FROM CONTINUOUS DISTRIBUTIONS

When the **Cumulative probability** option is exercised after one of the standard continuous distributions is selected with **Calc→Probability**, probabilities of the form $P(X \leq x_0)$ can be obtained. This means that **MINITAB** will produce the probability that the random variable X is no more than a given value x_0.

Conversely, **MINITAB** can generate a value x_0 such that $P(X < x_0) = p$ for a given value of p when the **Inverse cumulative probabilities** option is selected.

To see how this is accomplished, continue with the popcorn experiment in

Example 2. Enter the heading *Cum Prob* for column C3 and apply **Calc→Probability Distributions→Normal**. Change the settings from figure *Minitab 5.4* by selecting **Cumulative probability** and type *C3* in the **Optional Storage** box.

Column C3 will now contain values of $P(X \le x)$ for $x = 0, 1, 2, ..., 180$.

The following table of percentages of popped corn are obtained from the column of cumulative probabilities in C3 with the indicated times (in seconds) for this simulation:

Time	60	80	100	120	140	160	180
Percent popped	0%	6%	25%	59%	87%	98%	100%

Suppose that in order to avoid burning the popcorn, the microwave is to be turned off when 2% of the kernels remain unpopped. To determine the time when this occurs, select **Calc→Probability Distributions→Normal** and **Inverse cumulative probability**. Enter the information as described in figure *Minitab 5.6*.

Minitab 5.6

Observe from the above display that the input constant 0.98 has been entered in the **Input Constant** box. **MINITAB** will now generate the time x_0 such that $P(X \le x_0) = 0.98$. The displayed time in the **Session** window, is 159.1825.

One can conclude that 98% of the kernels will be popped after approximately 2 minutes and 39 seconds (159.1825 seconds), at which time the microwave should be turned off.

A team exploration problem in the **Lab #5:** *Data Sheet* has the student carry out this experiment with a microwave oven or other popping device. Depending on the power of the device and the brand of popcorn, one can expect varying results. **MINITAB** can be

utilized in producing a popping intensity graph based on the empirical evidence together with the graph of a normal curve approximation to the popping intensity.

4. NORMAL APPROXIMATIONS TO THE BINOMIAL DISTRIBUTION

A binomial random variable X has mean $\mu = np$ and standard deviation $\sigma = \sqrt{np(1-p)}$ where n is the number of trials and p is the probability of success for each trial. Suppose that X^* is a normal random variable having the same mean and standard deviation as X.

MINITAB can be utilized to demonstrate that under certain minimal conditions, such as $np > 5$ and $n(1-p) > 5$, the normal and binomial distributions have very similar shapes. Moreover, a probability of the form $P(X \leq x_0)$ from the binomial distribution can be demonstrated to be close to the probability $P(X^* \leq x_0 + 0.5)$ from the corresponding normal distribution. The add-on of 0.5 is commonly referred to as the **continuity correction factor** applied when converting from the discrete binomial distribution to the continuous normal distribution.

Example 3: A 1995 report from a Harvard University study claims that 34.9% of people of age 25 to 29 years are homeowners. Suppose that 50 of these people are randomly polled. Let X represent the number who are homeowners. The mean and standard deviation of X are as determined from the above formulas are $\mu = 17.45$ and $\sigma \approx 3.37$.

Let X^* denote the normal random variable having the same mean and standard deviation as X. **MINITAB** will be used to demonstrate how similar the shapes of the distribution of binomial probabilities and the normal densities are to each other. Comparisons of their probabilities will also be made.

First, note that virtually all of the responses should fall between $x = 7$ and $x = 28$ (within three standard deviations from the mean). Use **MINITAB** to store the integers from 7 to 28 in column C1 (labeled *x-values*) by applying **Calc→Set Patterned Data**. In **Store result in** enter C1; in **Start at** enter 7; in **End at** enter 28; in **Increment** enter 1.

Label column C2 as *P(X = x)* and apply **Calc→Probability Distributions →Binomial** using C1 as the input column and C2 as option storage. Enter 50 for the number of trials and 0.349 for the probability of success.

Label column C3 *Ndensity* and apply **Calc→Probability Distributions→Normal** to generate a table of normal densities of X* for x = 7, 8, ..., 28 in the manner described in figure *Minitab 5.7*.

```
┌─────────────────────────────────────────────────────────────────┐
│ ▬                    Normal Distribution                          │
├─────────────────────────────────────────────────────────────────┤
│  C1      x-values        ⊙ Probability density                    │
│  C2      P(X=x)          ○ Cumulative probability                 │
│  C3      Ndensity        ○ Inverse cumulative probability         │
│                                                                    │
│                          Mean:    [17.45    ]                      │
│                          Standard deviation:    [3.37    ]         │
│                                                                    │
│                          ⊙ Input column:         [C1     ]        │
│                             Optional storage:     [C3     ]        │
│                                                                    │
│                          ○ Input constant:        [       ]        │
│                             Optional storage:      [       ]        │
│       [  Select  ]                                                 │
│                                                                    │
│  [?] PDF                       [   OK   ]     [  Cancel  ]         │
└─────────────────────────────────────────────────────────────────┘
```

Minitab 5.7

In the **Data** window one will observe the integers 7 through 28 in column C1 (x-values), each of which represents the number who are homeowners out of the 50 people sampled. In column C2 (P(X=x)) are the binomial probabilities associated with the values x = 7, 8, ...28. In column C3 (Ndensity) are the values of the normal density function at the values x = 7, 8, ..., 28. Observe how close the column C3 values are to the C2 values.

Both the binomial probability distribution for X and the density function for X* are graphed by selecting **Graph→Plot** and filling the boxes as shown in figure *Minitab 5.8*. This is accomplished by clicking on **Frame→Multiple Graphs** and by selecting **Overlay graphs on the same page**.

Alternative headings on the x-axis and y-axis of *Number Of Homeowners* and *Probability* can be typed in on the graph after accessing **Frame→Axis→Label**. The title *Binomial vs Normal Distribution* is entered after **Annotation→Title**.

114

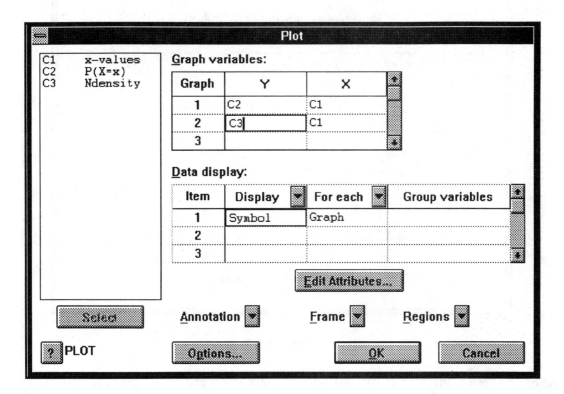

Minitab 5.8.

In the figure *Minitab 5.9* display of the two distributions, a circle represents a point on the binomial distribution and a plus for a point on the normal density function graph.

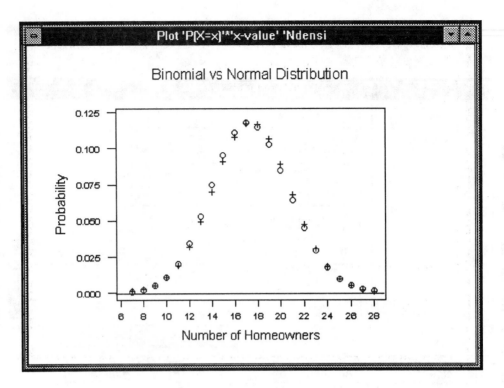

Minitab 5.9

As can be seen from this display, the normal distribution provides a good approximation to the binomial distribution in this example.

It was noted earlier that if X* is the normal approximation to a binomial random variable X having the same mean and standard deviation, then probabilities associated with X can be approximated with the corresponding probabilities using X* under most circumstances. If the number of trials is large, it may not be possible to evaluate a probability concerning X; in this event the normal approximation should provide a good estimate.

In *Example 3*, consider the probability that no more than 15 of the 50 people polled are homeowners. Then P(X ≤ 15) ≈ P(X* ≤ 15.5) which can be evaluated through **Calc→Probability Distributions→Normal** and by proceeding as indicated in figure *Minitab 5.10*.

```
┌─────────────────────────────────────────────────────────────┐
│ ▓▓                    Normal Distribution                     │
├─────────────────────────────────────────────────────────────┤
│ ┌──────────────┐      ○ Probability density                  │
│ │              │      ◉ Cumulative probability                │
│ │              │      ○ Inverse cumulative probability        │
│ │              │                                              │
│ │              │      Mean:     ┌─────────┐                   │
│ │              │                │ 17.45   │                   │
│ │              │      Standard deviation:  ┌─────────┐        │
│ │              │                           │ 3.37    │        │
│ │              │                                              │
│ │              │      ○ Input column:    ┌─────────┐          │
│ │              │                          │         │         │
│ │              │        Optional storage: ┌─────────┐         │
│ │              │                          │         │         │
│ │              │                                              │
│ │              │      ◉ Input constant:  ┌─────────┐          │
│ └──────────────┘                          │ 15.5    │         │
│  ┌──────────┐          Optional storage: ┌─────────┐          │
│  │  Select  │                             │         │         │
│  └──────────┘                                                 │
│  ┌─┐                          ┌──────────┐  ┌──────────┐      │
│  │?│ CDF                      │    OK    │  │  Cancel  │      │
│  └─┘                          └──────────┘  └──────────┘      │
└─────────────────────────────────────────────────────────────┘
```

Minitab 5.10

The **Session** window will display the value $P(X^* \leq 15.5) = 0.2814$.

The value of $P(X \leq 15)$ can be directly evaluated through the commands **Calc→ Probability Distributions→Binomial**. Enter *50* for the number of trials, *0.349* as the probability of success, and *15* in the **Input constant** box.

The **Session** window will display the value $P(X \leq 15) = 0.2851$ which is the probability that no more than 15 of the 50 polled people are homeowners. The normal approximation of 0.2814 in error by only 0.0037.

NOTES

LAB #5: *DATA SHEET*

Name: _____   *Date:* _____

Course #: _____   *Instructor:* _____

1. Suppose that the heights of trees in a forest are approximately normally distributed. The mean height of the trees is 24.8 feet and standard deviation is 8.4 feet. Let X represent the height of a randomly measured tree.

  (a) Enter heights *0, 1, 2, ..., 50* in column C1 (label: *x-values*), normal densities in column C2 (label: *Ndensity*), and cumulative probabilities in column C3 (label: *Cum Prob*). Provide a printout of these three columns from the **Data** window.

  (b) Use the commands **Stat→Plot** and select **Connect** from the **Display** option in order to generate a graph of the distribution of x-values from columns C1 and C2. Locate the points on the graph which are at the mean and at one and two standard deviations away from the mean.

  (c) Use the cumulative probabilities in column C3 to determine the percentage of trees which are:

  i. between 20 and 30 feet tall _____ .

  ii. greater than 35 feet tall _____ .

  Recall that for a < b,

$$P(a \leq X \leq b) = P(X \leq b) - P(X \leq a) \quad \text{and} \quad P(X > a) = 1 - P(X \leq a).$$

(d) Use the procedure described in figure *Minitab 5.6* involving inverse cumulative probabilities to find the first quartile Q_1, third quartile Q_3, 10th percentile P_{10}, and 90th percentile P_{90} of the heights of the trees of the forest. Also, locate these values on your display of the distribution of x-values obtained in part (c). Recall that for a given percentile P_k, at most k percent of the values are less than P_k.

$Q_1 = $ _____. $Q_3 = $ _____.

$P_{10} = $ _____. $P_{90} = $ _____.

(e) Use the methods described in *Example 1* for generating 200 random tree heights. Store these heights in column C4 and label as *Heights*. Find the sample mean $\bar{x}$ and sample standard deviation s of the heights of these 200 trees using **Stat→Basic Statistics→Descriptive Statistics**. Compare with the population mean μ and population standard deviation σ.

$\bar{x} = $ _____. $s = $ _____.

(f) Display a frequency histogram of the heights of these 200 trees over the interval [0, 50] in the manner shown in figure *Minitab 5.3*. In **Options**, select **CutPoint**; start at 0, end at 50, and choose an increment of 5. Provide proper labels and include data labels. Compare the shape of this histogram with the distribution of the population of trees from part (a).

(g) Find the percentage of these 200 heights which are between 20 and 30 feet tall; greater than 35 feet tall. Compare these percentages with those found in part (c).

(h) Use **MINITAB** to sort the 200 tree heights from smallest to largest by using **Manip→Sort**. Store the sorted values in column C5. Use this information to find the percentage of the tree heights which fall within two standard deviations of the mean. Compare your answer with the statement of the **Empirical Rule**.

Interval of values within two standard deviations of the mean:
_____ to _____

Number and percentage of the 200 heights in this interval _____.

2. Suppose that the lifetime X of a randomly tested light bulb describes an exponential distribution having a mean life of 1000 hours. A tested light bulb is turned on until it fails. Recall that an exponential distribution is positively skewed for $x \geq 0$ and has a density function defined by

$$f(x) = \begin{cases} 0 & \text{if } x < 0 \\ \left(\dfrac{1}{\mu}\right) e^{-x/\mu} & \text{if } x \geq 0 \end{cases}$$

The standard deviation σ is equal to the mean μ.

(a) Use the methods described in figure *Minitab 5.4* and figure *Minitab 5.5* to display a graph of the density function for X over the interval of x-values [0, 4000]. Apply **Calc→Set Patterned Data** to store 0, 50, 100, 150, ..., 4000 in column C1 with heading *Lifetime*. Then select **Calc→Probability Distributions →Exponential** to generate the density function values from the C1 column into column C2 with heading *Density*. Also, store the cumulative probabilities for these x-values in column C3 with heading *Cum Prob*. Provide a printout of these three columns from the **Data** window. Locate the mean on your graph.

(b) How would you describe the shape of the distribution of the population of bulb lifetimes from your graph in part (a)? Use answers such as bell-shaped, skewed to the right, etc.

(c) Use the cumulative probabilities from column C3 to determine the percentage of bulbs

   i. that are expected to fail before 500 hours _____ ;

   ii. that are expected to fail after 2000 hours _____ .

(d) Use the procedure described in figure *Minitab 5.6* involving inverse cumulative probabilities to find the first quartile Q_1, median Q_2, and third quartile Q_3. Locate these on your display of the distribution of lifetimes given in part (a).

First quartile Q_1 = _____ .

Median Q_2 = _____ .

Third quartile Q_3 = _____ .

(e) Using the method described in figure *Minitab 5.2* and figure *Minitab 5.3*, simulate the random testing of 500 light bulbs to determine their lifetimes. Display a histogram over the interval [0, 4000] using an increment of 500. Provide data labels. Compare the shape of your histogram with the probability distribution displayed in (a).

(f) Based on your histogram in (c), find the percentage of these light bulb which lasted fewer than 500 hours; more than 2000 hours. Compare your answers with the expected percentages found in part (c).

(g) Apply **Stat→Basic Statistics→Descriptive Statistics** to the 500 bulb lifetimes so as to determine the sample mean and sample standard deviation. Compare these with the corresponding population mean and population standard deviation. Also, give the values of the quartiles and median and compare with your answers from part (d).

$\bar{x}$ = _____ .   s = _____ .   Median = _____ .

First quartile Q_1 = _____ .   Third Quartile Q_3 = _____ .

(h) Position the mean, median and quartiles on the horizontal axis of your histogram in part (c). Also, estimate on your graph where the mode is to be located. Discuss the relative positions of the mean, median, and mode. Is this what you would expect for a distribution of this shape?

3. According to a 1995 International Data Corp. / AP report, 67% of PC operating systems use Microsoft's *Windows*. Suppose that 500 owners of a personal computer system are polled. Let X be the binomial random variable which counts the number who use a *Windows* operating system.

(a) Determine the mean μ and standard deviation σ of X.

(b) Let X* be a normal random variable which has the same mean and standard deviation as X. Apply the procedure explained in *Example 3* and figure **Minitab 5.10** to estimate the probability that fewer than 300 of the 500 operating systems use *Windows*; that at least 350 use *Windows*.

(c) Determine the x-values for this binomial random variable which are within three standard deviations of the mean. Recall that $\mu = np$ and $\sigma = \sqrt{np(1-p)}$. Apply **Calc→Set Patterned Data** to store these in column C1 with heading *x-value*. Generate binomial probabilities in column C2 with heading *P(X=x)*. In column C3 (*Ndensity*) store the density function values from C1 for the normal approximation X* to X. As demonstrated in figure **Minitab 5.9**, construct a graph of the probabilities in column C2 and the densities in column C3. Based on this display, compare the two distributions.

x-values _____.

4. **Team Exploration Project.** Recall the popcorn popping problem discussed in *Example 2*. Carry out the following procedures and type a detailed report explaining your observations and conclusions.

(a) Set up your own experiment of this popcorn problem. Decide on what kind of popper to use (microwave or otherwise) as well as the brand of popcorn. By popping several bags of the popcorn, estimate the value of the mean and standard deviation for a normal distribution of popping times.

(b) Use the methods of figure *Minitab 5.4* and figure *Minitab 5.5* to display a printout of a graph of the distribution of popping times. Provide appropriate labels. Also, display a table of percentages of popcorn popped after each 10 seconds of elapsed time.

(c) Suppose that the popping procedure used is to stop when 2% of the kernels remain unpopped. Using the procedure from figure *Minitab 5.6*, find the theoretical time when popping should stop. Also, explain how well this compares with your experimental results.

(d) Explain whether or not you believe the normal distribution model is appropriate for this experiment.

NOTES

STATISTICS LAB # 6

SAMPLING DISTRIBUTIONS AND THE CENTRAL LIMIT THEOREM

PURPOSE - to use MINITAB to

1. understand **randomness** or **chance variation**
2. investigate the **sampling distribution** of the **sample means**
3. apply the **Central Limit Theorem (CLT)**

BACKGROUND INFORMATION

1. **Random Sample** - a sample obtained in such a way that each of the possible samples of fixed size (n) has an equal probability of being selected.

2. **Sampling Distribution of a Sample Statistic** - the distribution of values for that sample statistic obtained from all possible samples of a population. The sample sizes must all be the same, and the sample statistic could be any descriptive sample statistic.

3. **Central Limit Theorem** - if a random sample of size n is selected from any population with mean μ and standard deviation σ, then the distribution of values from $Z = \dfrac{\overline{X} - \mu}{\sigma / \sqrt{n}}$, where $\overline{X}$ is the sample mean, tends to the standard normal distribution as $n \to \infty$. That is, Z is approximately normally distributed for large n.

4. **Sampling Distribution of the Sample Mean** - if all possible random samples of size n are taken from any population with mean μ and standard deviation σ, then the *sampling distribution of the sample mean* will have

   (a) mean μ

   (b) standard deviation $\sigma / \sqrt{n}$

   (c) a normal distribution when the sampled population is normally distributed or will be approximately normally distributed for samples of size 30 or more when the sampled population is not normally distributed. *Note: for some distributions with n < 30, $\overline{X}$ could be approximately normally distributed.*

4. **Standard Error of the Mean** - the standard deviation of the sampling distribution of sample means given by $\sigma / \sqrt{n}$.

5. **Normal Probability Plot** - this is a graphical technique for examining whether the distribution of a random variable is normal. If the distribution is normal, this plot should have a straight-line appearance. Substantial departures from the straight-line appearance indicate a violation of the normality assumption. This graphical technique will be used to help determine whether the distribution of the sample means is normal.

PROCEDURES

Load the **MINITAB** (windows version) software as described in *Lab #0*.

1. GENERATING RANDOM SAMPLES, COMPUTING SAMPLE MEANS AND INVESTIGATING THE DISTRIBUTION OF THE SAMPLE MEANS

Example 1: Use **MINITAB** to simulate the rolling of a regular six-sided die. Roll the die 50 times and repeat the experiment 200 times. We will use this generated data to record the sample means, and to investigate the distribution of these 200 sample means. (*Note: If you have the student version adjust the number of rolls and repetition appropriately*).

First, to simulate the rolling of the die, use the mouse to select **Calc** and a drop down box will appear. In that drop down box select **Random Data** . The two drop down boxes are shown in figure *Minitab 6.1*. The different population distributions available in **MINITAB** are listed in the **Random Data** drop down box.

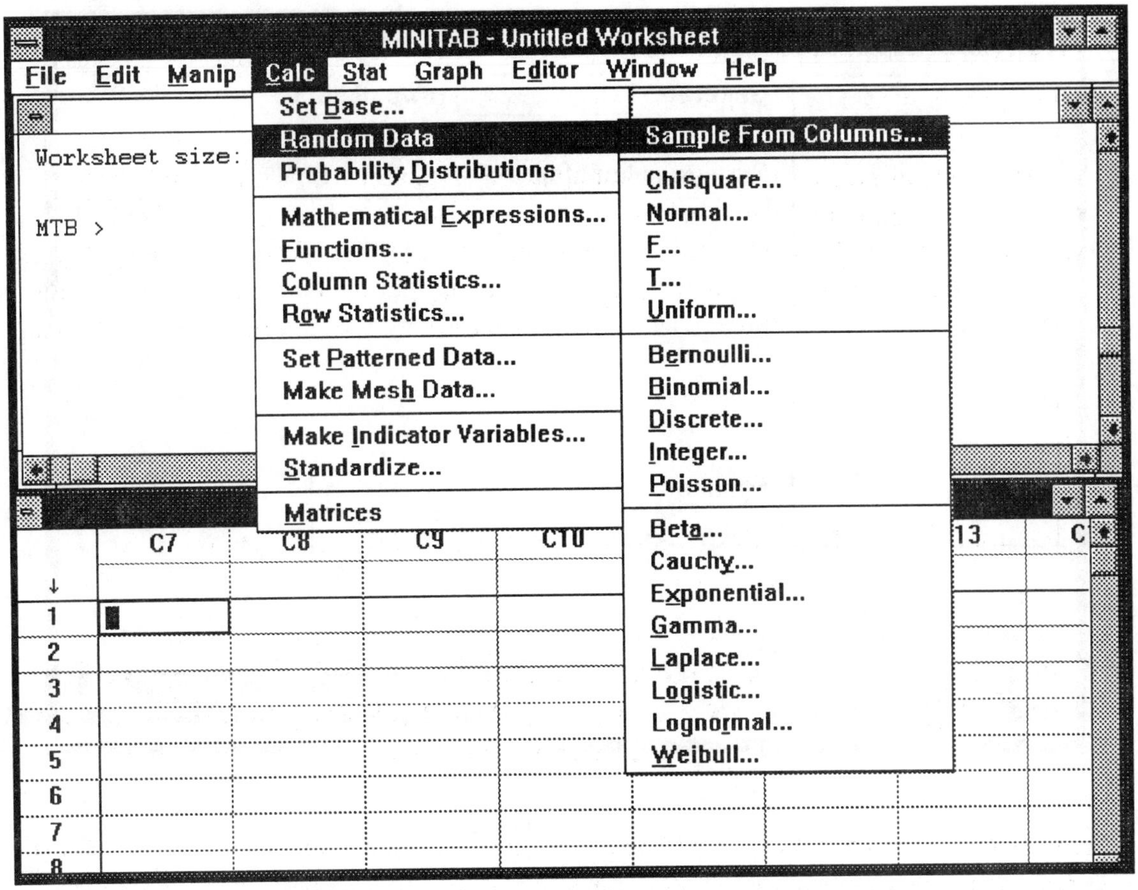

Minitab 6.1

Use the mouse to select the **Integer** option. The **Integer Distribution** (discrete uniform) dialog box will be displayed. At the **Generate** prompt, type *200* in the text box. Use the mouse to select the **Store in Column(s)** box and type *C1 - C50* in the text box. Use the mouse to select the **Minimum value** box and type the value *1* in the text box. Use the mouse to select the **Maximum value** box and type the value *6* in the text box. Minimum value of 1 and maximum value of 6 are used since the die is a regular six-sided die and we are assuming that the faces will be numbered 1 to 6. The **Integer Distribution** dialog box is shown below in figure *Minitab 6.2* with the appropriate entries.

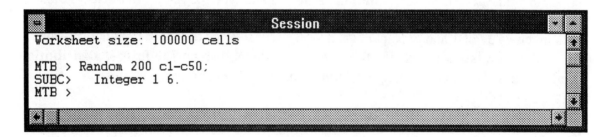

Minitab 6.2

Next, use the mouse to select the **OK** button to tell **MINITAB** to generate the data
set. The preceding instructions will enable **MINITAB** to generate 50 columns (C1 - C50)
of 200 rolls each of a fair 6-sided die. *Note: each row of 50 values represents a random
sample of 50 rolls of the die.* The equivalent **MINITAB** commands are displayed in the
following **Session** window as shown in figure *Minitab 6.3*. This window is obtained by
clicking on the **Session** window.

```
                              Session
Worksheet size: 100000 cells

MTB > Random 200 c1-c50;
SUBC>    Integer 1 6.
MTB >
```

Minitab 6.3

You can use the mouse to click on the **Data** window to observe the generated data.
You will have to scroll the screen down as well as to the right to observe all the generated
data. Also, you can use the mouse to select **File→Display Data**. In the **Display Data**
dialog box, use the mouse to select C1 through C50 by dragging the arrow from C1 to
C50 in the left hand side box and then clicking on the **Select** button. The columns

C1- C50 will be displayed in the right hand side box. Alternatively, you can type *C1 - C50* in the **Display Data** dialog box as shown in figure *Minitab 6.4*.

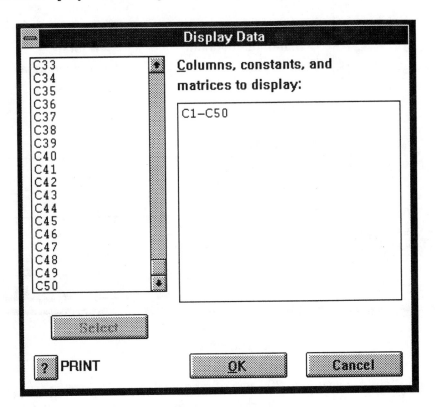

Minitab 6.4

Select the **OK** button and the data will be generated in the **Data** window.

Next, use the mouse to select **Calc→Row Statistics** and select **Mean**. Suppose we want to consider samples of size 25. In the **Input variable box** select C1- C25 (say) with the mouse or type in *C1- C25*. In the **Store result** box type in *C51*. The **Row Statistics** dialog box should look like the one in figure *Minitab 6.5* with **Mean** highlighted.

Minitab 6.5

This is equivalent to generating 200 sets of random samples of size 25 from the set {1, 2, 3, 4, 5, 6}. Also, C51 stores the means of these 200 sets of random samples of size 25 generated from the ***discrete uniform*** distribution.

Next, select **Graph→Histogram**. In the **Graph variables** box for graph 1, select C51 and select **OK**. A histogram of the average values should be displayed in a **Histogram** window. Observe the general shape of the histogram.

Exercise 1: (See Problem 3 in the **Data Sheet**). Repeat the above process of computing the means for C1- C2, C1- C10, C1- C20, C1- C30, C1- C40, C1- C50 and save in columns C52, C53, C54, C55, C56 and C57. Draw histograms for these computed column means and observe their general shape. ***These graphs should give you insight into the sampling distribution of the sample mean for different sample sizes (2, 10, 20, 30, 40 and 50) for this discrete uniform distribution.*** In addition, generate descriptive statistics values for the sample means in C51 - C57 by selecting **Stat→Basic Statistics→ Descriptive Statistics**. In the **Variable** text box type in *C51 - C57* and select **OK**. Observe the sample means and compare to the ***population mean of 3.5***. Also, the ***population variance is 2.9167***. *You should try to verify these two values.* Compare the standard deviations for the column means for the different sample sizes (obtained with the **Descriptive Statistics** procedure) with the standard error of the means. Next, apply the **Empirical Rule** to these column means. ***Hint:*** You should sort these column means in ascending order using **Manip → Sort**; before you estimate what proportions are between one standard deviation, two standard deviations, and three standard deviations from the mean. Problem 6 in the **Data Sheet** provides further information on this procedure. To estimate the proportions, count how many sample means are within one standard deviation from the column mean and divide by the number of column values (200 in this case), etc.

Before attempting Example 2, erase the data in the occupied columns for Example 1.

NOTE: To delete a list of columns, one can choose **Manip→Erase Variables**. This command allows you to delete any combination of columns, stored constants, and matrices. The **Erase Variables** dialog box is shown below in figure *Minitab 6.6*. The last column is represented by C? - you need to type in the column number.

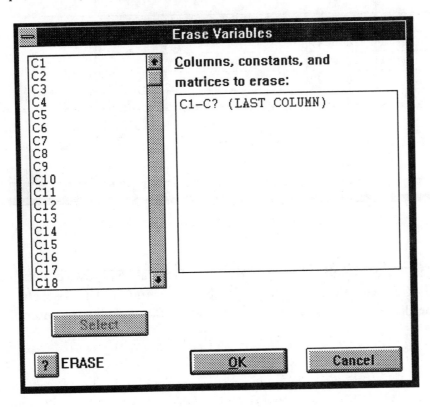

Minitab 6.6

Example 2: Generate 500 values from the following discrete distribution. Repeat the experiment 50 times and store in columns C3-C52. *(Note: If you have the student version adjust the number of rolls and repetition appropriately).*

x	P(X = x)
1	0.0249
2	0.0741
3	0.2132
4	0.3621
5	0.3257

Enter the above values in C1 and C2. The **Data** window with these values is shown below in figure *Minitab 6.7*. Note that this discrete distribution is skewed to the left.

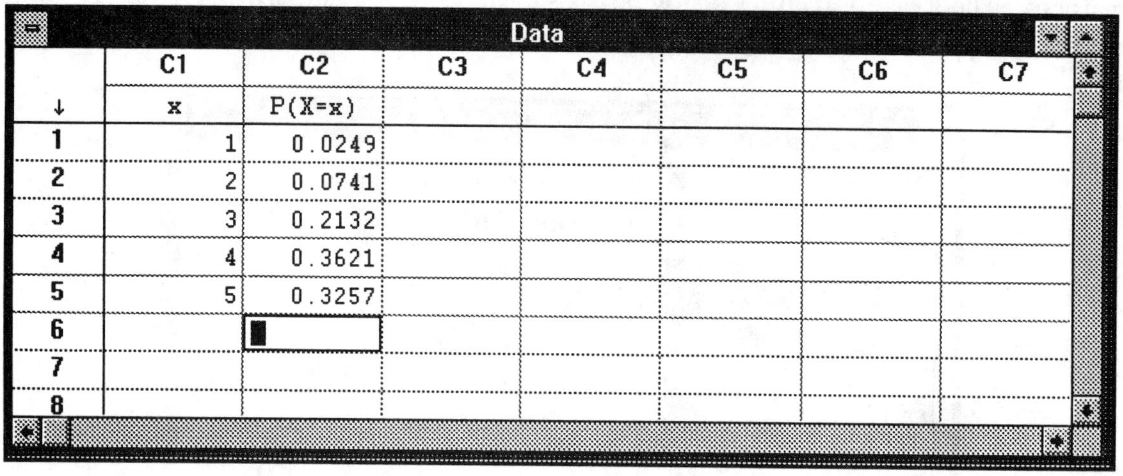

	C1	C2	C3	C4	C5	C6	C7
↓	x	P(X=x)					
1	1	0.0249					
2	2	0.0741					
3	3	0.2132					
4	4	0.3621					
5	5	0.3257					
6							
7							
8							

Minitab 6.7

Next select **Calc→Random Data→Discrete**. Generate *500* rows of data and store in columns *C3 - C52* with **Values in** *C1* and **Probabilities in** *C2*. The **Discrete Distribution** dialog box is shown below in figure *Minitab 6.8* with the appropriate entries.

```
┌─────────────────────────────────────────────────────────────┐
│ ▬            Discrete Distribution                          │
├─────────────────────────────────────────────────────────────┤
│ C1    x          Generate  │500        │  rows of data       │
│ C2    P(x)                                                   │
│                  Store in column(s):                        │
│                  ┌────────────────────────────────────────┐ │
│                  │ c3-c52                                 │ │
│                  │                                        │ │
│                  │                                        │ │
│                  └────────────────────────────────────────┘ │
│                                                             │
│                  Values in:         │c1        │            │
│                  Probabilities in:  │c2        │            │
│                                                             │
│              ┌──────────┐                                   │
│              │  Select  │                                   │
│              └──────────┘                                   │
│   ┌─┐                                                       │
│   │?│ RANDOM          ┌───OK───┐   ┌──Cancel──┐            │
│   └─┘                 └────────┘   └──────────┘            │
└─────────────────────────────────────────────────────────────┘
```

Minitab 6.8

Note: Each row represents a random sample of size 50 from the sampled population.

Exercise 2: (See Problem 4 in the **Data Sheet**). Select a few of the generated columns (C3 - C52) and use the **Graph** option to display histograms of sample means from these columns. They should be skewed to the left. Next, compute row means for C3 - C10, C3 - C20, C3 - C30, C3 - C40, and C3 - C50 by using the **Calc** option and place in columns C53, C54, C55, C56, and C57. Repeat the requirements for *Exercise 1.*

2. VERIFYING THE DISTRIBUTION OF THE SAMPLE MEANS BY USE OF NORMAL PROBABILITY PLOTS (NPP)

To display *normal probability plots* for column *C53* in *Example 2*, the computed means for the 500 samples of size 8 (C3 - C10), select **Graph→Normal Plots→ Variable (C53)**. The **Normal Probability Plot** dialog box will appear and is shown below in figure *Minitab 6.9* with the appropriate entries.

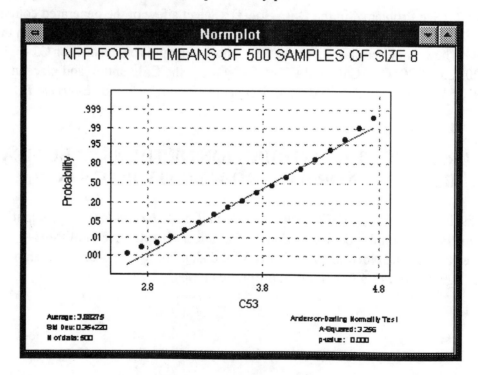

Minitab 6.9

When the graph is drawn, *if the sample came from a normal distribution, the plot will be approximately linear. If the sample is from a non-normal population, the plot usually shows some curvature.* A sample NPP is shown below for the means of the 500 samples of size eight (C53) from the discrete distribution in *Example 2. Observe that the plot is non-linear indicating that the 500 sample means did not come from a normally distributed population.* This is not surprising since the sample size is very small (n = 8). Figure *Minitab 6.10* shows the normal probability plot.

Minitab 6.10

136

NOTE: Later in your course work you will do hypothesis tests that will enable you to interpret the *p-value* given in the **Normplot** dialog box. In figure *Minitab 6.9*, the test for normality selected was the **Anderson-Darling** test. This will test whether the sample means came from a normal population or distribution. A very small p-value (in this example, the p-value = 0) will indicate that the distribution is not normal.

Exercise 3: (See Problem 5 in the **Data Sheet**). For the means generated for different sample sizes from the discrete distribution in *Example 2*, graph the NPPs and analyze.

NOTES

LAB #6: *DATA SHEET*

Name: _____   *Date:* _____

Course #: _____   *Instructor:* _____

1. (a) Fill out the distribution table below for the discrete uniform distribution in *Example 1.*

x						
P(x)						

   (b) Compute the population mean μ for the distribution in *Example 1.* (See *Lab #4* for formulas).

   $\mu =$ _____.

   (c) Compute the population standard deviation σ for the distribution in *Example 1.*

   $\sigma =$ _____.

2. Compute the mean μ and standard deviation σ for the population distribution given in *Example 2.*

   $\mu =$ _____.

   $\sigma =$ _____.

3. Relate your answers in Problem 1 to your observations after working *Exercise 1.* Present your results in a report. Include all necessary graphs, appropriate tables with computations, etc. Make sure you discuss your observations for each component of the exercise.

4. Relate your answers in Problem 2 to your observations after working *Exercise 2*. Present your results in a report. Include all necessary graphs, appropriate tables with computations, etc. Make sure you discuss your observations for each component of the exercise.

5. Discuss your observations after working *Exercise 3* and relate them to the sampling distribution of the sample means. Present your observations in a report with the necessary NPP plots.

6. Suppose that a certain brand of 40-watt bulb advertises that the mean life of the bulb is 1000 hours. Assume that the lifetimes of the bulbs are *exponentially distributed* (so the standard deviation is also 1000 hours). Use **MINITAB** to generate 200 random samples, each having size 50, from this population. You will be creating columns C1- C50, where each column has 200 entries. *Hint:* Select **Calc→Random Data→Exponential**.

Present your graphs, observations, calculations etc. in a report. You should use tables to present the results of your computations for the different columns.

(a) Produce a histogram for column C1. Describe the shape of the distribution of C1 values. Does this shape represent the distribution of the lifetimes of the bulbs? Discuss.

(b) Generate the 200 sample means using columns C1 - C10, C1 - C20, C1 - C30, C1 - C40, C1 - C50 and store in C51 - C55 as in previous exercises. Produce histograms for the sample means for each column from C51 - C55. Also, find the mean and standard deviation for each column from C51 - C55. Compare with what you would expect based on the distribution of the sample means.

(c) Sort the values of C51 (here the sample size n = 10) using **MINITAB** by selecting **Manip→Sort**. In the **Sort Column** text box type *C51*. In the **Store sorted column(s) in** text box type *C56*. In the **Sort by Column** text box type *C51* and select the **OK** button.

   Using the values in C56, estimate the probability that a sample mean of 10 light bulbs selected has a lifetime of between 900 and 1000 hours. Explain your logic as to the way you got your answer.

(d) Repeat part (c) for columns C52 - C55. You may store the sorted values in C57 - C61.

(e) Work the probability problem in parts (c) and (d) by using the methods of your text (normal distribution table). Compare your answers. **Show your work for C51 (C56).**

7. **Team Exploration Project**. The two examples in this *Lab* dealt with discrete distributions. There are several continuous distributions with the *Normal* being the most commonly used. From the **Calc→Random Data→Normal** selection sequence, generate 500 samples of size 10 from a normal distribution with mean $\mu = 21$ and standard deviation $\sigma = 4$. Select sample sizes of 2 to 10, compute their means, and store in appropriate columns. Use **Stat→Basic Statistics→ Descriptive Statistics** sequence to compute the means and standard deviations for these columns. Compare these column means with the population mean, and these standard deviations with the standard errors. Draw histograms and NPPs for the generated column means. Discuss your results in a report. Are there any differences between your observations here and with the discrete distributions given in Problems 3 and 4? Explain. (*Note: If you are using the student version, you will need to adjust the number of samples to generate*).

NOTES

STATISTICS LAB # 7

UNBIASED AND BIASED ESTIMATORS OF POPULATION PARAMETERS

PURPOSE - to use MINITAB to

1. explore the relationship between a **statistic** and the corresponding **parameter**
2. simulate random samples from various populations, such as the **normal**, **Student's t**, and **chi-square distribution,** using both small and large sample sizes
3. obtain a better understanding of those statistics which are **unbiased estimators** and those which are **biased** estimators of their related parameters
4. compare the **empirical probabilities** obtained by random sampling with the corresponding **theoretical probabilities**

BACKGROUND INFORMATION

1. **Parameter -** a numerical characteristic of a population, such as population mean μ and variance σ^2.

2. **Statistic -** a numerical characteristic of a sample, such as sample mean $\bar{x}$ and sample variance s^2.

3. **Unbiased estimator of a parameter -** a sample statistic whose expected value or mean equals the value of the parameter it is estimating. For example, the sample mean $\bar{X}$ is an unbiased estimator of the population mean μ in that $\mu_{\bar{x}} = E(\bar{X}) = \mu$ where $E(\bar{X})$ is the expected value (mean value) of the distribution of $\bar{x}$ s.

4. **Biased estimator of a parameter -** a sample statistic whose expected value (mean) is not equal to the parameter it is estimating.

5. **Student's t distribution -** the probability distribution of the Student's t-statistic having degrees of freedom is $df = n - 1$ given by the equation $t = \dfrac{\bar{x} - \mu}{s/\sqrt{n}}$. The mean of this distribution is zero and the standard deviation has a value greater than one.

6. **Chi Square distribution -** the probability distribution of the chi-square statistic having degrees of freedom $df = n-1$ is given by the equation $\chi^2 = \dfrac{(n-1)s^2}{\sigma^2}$.

143

PROCEDURES

In *Lab #6*, the **Central Limit Theorem** was investigated through various activities involving **MINITAB**. The mean of the sampling distribution of sample means is equal to the population mean μ. Other parameters and statistics which estimate them will now be explored.

1. GENERATING RANDOM SAMPLES OF STANDARD DEVIATIONS AND VARIANCES FROM A POPULATION

Random samples of a fixed size can be generated from both discrete and continuous distributions with **MINITAB** by selecting **Calc→Random Data**.

Example 1: Consider the experiment of tossing a fair coin once and recording whether a head or a tail shows. By coding a head as 1 and a tail as 0, the resulting population is Bernoulli with p = 0.5, which is the probability of a heads appearing. **MINITAB** will now simulate this experiment a total of 10 times and then calculate the values of the sample standard deviation s and variance s^2. To explore whether S and S^2 are unbiased estimators of σ and σ^2, respectively, this process will be carried out a total of 300 times.

After selecting **Calc→Random Data→Bernoulli**, fill in the **Bernoulli Distribution** dialog box as shown in figure *Minitab 7.1*.

Minitab 7.1

144

The **Data** window will now contain 10 columns C1 through C10 consisting of 300 rows of zeros and ones. Each row of 10 values can be viewed as a random sample selected from the population. We will now explore the distributions of these 300 sample standard deviations and variances.

To generate the sample standard deviations s in column C11 (label: *St Dev*), select **Calc→Row Statistics**, click on the **Standard deviation** circle, type *C1-C10* in the **Input variables** text box, and type *C11* in the **Store results in** text box.

To produce the sample variances s^2 in column C12 (label: *Variance*), select **Calc→Mathematical Expressions**, and type *C12* in the **Variable** and *C11**2* in the **Expression** text boxes, respectively.

The mean and standard deviation of the 300 standard deviations in column C11 and variances in column C12 are obtained by selecting **Stat→Basic Statistics →Descriptive Statistics** and typing *C11-C12* in the **Variables** text box. Figure *Minitab 7.2* displays these statistics when this simulation is performed once.
Note: the results will be different due to randomness when this simulation is performed again.

```
                                    Session
MTB > Random 300 C1-C10;
SUBC>     Bernoulli 0.5.
MTB > RStDev C1-C10 C11.
MTB > Let C12 = C11**2
MTB > Describe C11-C12.


                  N       MEAN     MEDIAN     TRMEAN      STDEV     SEMEAN
St Dev          300    0.49790    0.51640    0.50481    0.04935    0.00285
Variance        300    0.25033    0.26667    0.25551    0.03863    0.00223

                MIN        MAX         Q1         Q3
St Dev      0.00000    0.52705    0.48305    0.51640
Variance    0.00000    0.27778    0.23333    0.26667

MTB >
```

Minitab 7.2

Observe that the mean of the 300 sample variances is 0.25033, which is very close to the variance of the population given by $\sigma^2 = p(1 - p) = 0.25$. For this particular simulation, it may appear that s^2 is an unbiased estimator of σ^2. However, only one run of this simulation is insufficient to draw this conclusion! If this simulation were repeated a large number of times, it is likely that the sample variances obtained will mostly be close in value to the population variance $\sigma^2 = 0.25$.

From figure *Minitab 7.2* we note in this simulation that the mean of the 300 sample standard deviations is 0.49790 which is close to the population standard deviation of $\sigma = 0.5$. In this simulation there is nearly a 1% error in the standard deviation approximation and a 0.1 % error in the variance approximation. Again, further runs of this simulation are required in order to base an opinion of the bias or non-bias of the statistic S of the parameter σ.

While the mean of the distribution of sample variances of a fixed sample size n equals the population variance, the mean of the distribution of sample standard deviations is only approximately equally equal to the population standard deviation. This approximation becomes much better as n increases in value. To summarize,

(a) **the sample variance S^2 is an unbiased estimator of σ^2, i.e., $\mu_{S^2} = \sigma^2$;**

(b) **the sample standard deviation s is not an unbiased estimator of σ, but approaches being unbiased for large value of n, i.e., $\mu_S \approx \sigma$.**

It is of interest to display graphs of the distributions of these 300 standard deviations and 300 variances so as to make observations about the shapes of the distributions of standard deviations for sample size of 10. For the standard deviations in column C11, select **Graph→Histogram**, type *C11* below **X**, click on **Display data labels** in **Annotation→Data Labels**, and type *Distribution Of 300 Standard Deviations* in the **Annotation→Title** text box. As seen from figure *Minitab 7.2*, the minimum and maximum of the 300 deviations are 0.0 and 0.52705, respectively. For a histogram with six classes, select **Options**, click on the **Cutpoint** box and type in 0.0 : 0.54 / 0.09 in the **Define intervals using values** text box. As a consequence of carrying out this procedure, figure *Minitab 7.3* displays a histogram of the 300 standard deviations.

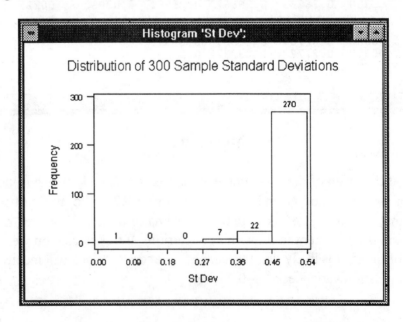

Minitab 7.3: Coin Toss Experiment - Standard Deviations

146

Observe that the distribution of these 300 sample standard are skewed to the left. Another way of drawing this conclusion is to note that the mean is less than the median. The mean and median for this simulation are found in figure *Minitab 7.3* to be 0.25033 and 0.26667, respectively.

As in the procedure discussed for the sample standard deviations, one can display a histogram for the 300 sample variances from column C12. Select **Graph→Histogram**, type *C12* below **X**, click on **Display data labels** in **Annotation→Data Labels**, and type *Distribution Of 300 Variances* in the **Annotation→Title** text box.

From figure *Minitab 7.2* one observes that the minimum and maximum of the 300 variances are 0.0 and 0.27778, respectively. For a histogram with seven classes, select **Options**, click on the **Cutpoint** box and type in *0.0 : 0.28 / 0.04* in the **Define intervals using values** text box. See figure *Minitab 7.4* for a display a histogram of the 300 standard deviations.

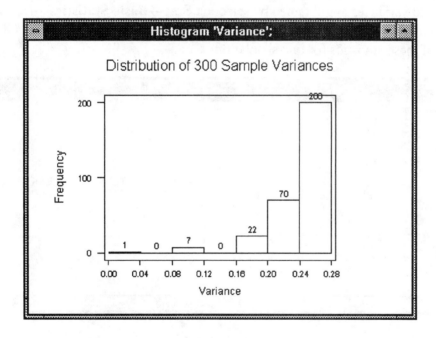

Minitab 7.4: Coin Toss Experiment - Variances

As with the sample standard deviations, it appears that the data is skewed to the left. This can also be verified by noting from figure *Minitab 7.2* that the sample mean of 0.25033 is smaller than the sample median of 0.26667.

Example 2: Consider random samples of size three selected from the standard normal population, which has mean $\mu = 0$ and standard deviation $\sigma = 1$. As in *Example 1*, the distribution of sample standard deviations and sample variances will be investigated using **MINITAB** to see if it is reasonable to conclude that S^2 is an unbiased estimator of σ^2 and S is a biased estimator of σ. For this purpose, 500 such standard deviations and variances will be generated from samples of size four.

147

Select **Calc→Random Data→Normal** and fill in the dialog boxes with 500 for the number of rows of data, *C1-C3* in **Store in column(s)**, *0* in **Mean**, and *1* in **Standard deviation**. See figure *Minitab 7.1* for a similar dialog box. There will now be 500 rows each consisting of random samples of size three selected from a standard normal population.

To generate the 500 sample standard deviations s from this data, select **Calc→Row Statistics**, click on the **Standard deviation** circle, type *C1-C3* in the **Input variables** text box, and type *C4* in the **Store results in** text box. Label C4 as *St Dev*.

To produce the 500 sample variances s^2 in column C5 (label: *Variance*), select **Calc→Mathematical Expressions**, type *C5* in the **Variable** text box, and type *C4**2* in the **Expression** text box.

The mean and standard deviation and other statistics of the 500 sample standard deviations and variances are obtained by selecting **Stat→Basic Statistics→Descriptive Statistics** and by typing *C4-C5* in the **Variables** text box. Figure *Minitab 7.5* described the values of these statistics for one simulation.

```
┌─────────────────────────── Session ───────────────────────────┐
│ MTB > Random 500 C1-C3;                                        │
│ SUBC>   Normal 0.0 1.0.                                        │
│ MTB > RStDev C1-C3 C4.                                         │
│ MTB > Let C5 = C4**2                                           │
│ MTB > Describe C4-C5.                                          │
│                                                               │
│                 N      MEAN    MEDIAN    TRMEAN    STDEV   SEMEAN │
│ St Dev        500    0.8993    0.8694    0.8812   0.4450   0.0199 │
│ Variance      500    1.0064    0.7559    0.9013   0.9506   0.0425 │
│                                                               │
│                MIN      MAX       Q1        Q3                  │
│ St Dev      0.0504   2.5359   0.5672    1.1833                  │
│ Variance    0.0025   6.4306   0.3217    1.4003                  │
│                                                               │
│ MTB > |                                                       │
└───────────────────────────────────────────────────────────────┘
```

Minitab 7.5

Note that the mean of these 500 sample variances in column C5 equals 1.0064 which is very close to the population variance $\sigma^2 = 1$ for the standard normal distribution. As in *Example 1* this tends to substantiate the fact that S^2 is an unbiased estimator of σ^2.

The mean of the 500 sample standard deviations in column C4 equals 0.8993 which is approximately 10% less than the population standard deviation $\sigma = 1$. This can be expected since S is an unbiased estimator of σ and the sample size n = 3 is very small.

Note: In any simulation of the above type, the values you obtain will differ from the examples discussed above.

As in *Example 1*, displays of the distributions of the 500 standard deviations and variances can be accomplished by producing two histograms by following the procedure discussed earlier.

Figures *Minitab 7.6* and *Minitab 7.7* shown below demonstrate that the distributions of the variances and standard deviations from random samples of size three selected from the standard normal population are skewed to the right.

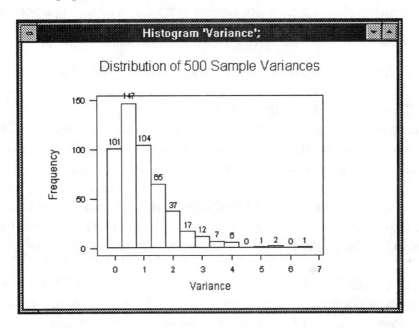

Minitab 7.6: Standard Normal Population - Variances

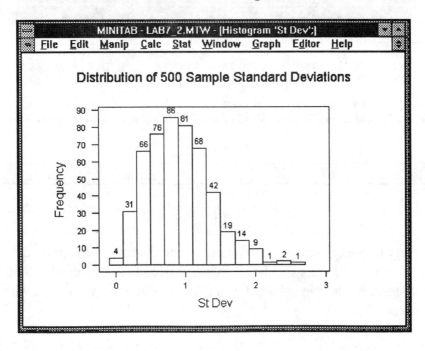

Minitab 7.7: Standard Normal Population - Standard Deviations

Further explorations relating to the distribution of sample variances and sample standard deviations will be encountered in the **Lab #7:** *Data Sheet*.

2. SIMULATING THE CHI-SQUARE DISTRIBUTION

When random samples of size n are generated from a normal population, the values of the statistic $\chi^2 = \dfrac{(n-1)S^2}{\sigma^2}$ will have a chi-square distribution with df = n-1 degrees of freedom. This can be investigated through simulation using **MINITAB**.

Example 2 (Continued): Recall that in the random selection of 500 samples of size three from a standard normal population, the variances are located in column C5. The 500 χ^2-values will be created in column C6 (label: *Chi Sqr*) by noting from the above formula that $\chi^2 = 2S^2$ since there are two degrees of freedom and $\sigma = 1$.

First, select **Calc→Mathematical Expressions**. In the **Variable** text box type *C6* and in the in the **Expression** text box type *2∗C5*. After clicking on **OK**, column C6 will consist of 500 chi-square values.

Select **Stat→Basic Statistics→Descriptive Statistics** and type *C6* in the **Variables** text box in order to display statistical information about these chi-square values.   Figure *Minitab 7.8* shows a display of the **Session** window.

```
                            Session
MTB > Let C6 = 2*C5
MTB > Describe  'Chi Sqr'.

                 N      MEAN    MEDIAN    TRMEAN     STDEV    SEMEAN
Chi Sqr        500    2.0128    1.5118    1.8027    1.9011    0.0850

                MIN       MAX        Q1        Q3
Chi Sqr      0.0051   12.8612    0.6434    2.8006

MTB > |
```

Minitab 7.8

One property of the chi-square distribution having degrees of freedom df = n-1 is that the mean is $\mu = n - 1$. In this simulation, observe from figure *Minitab 7.8* that the mean of the 500 chi-square values in column C6 is 2.0128 is in close agreement with the population mean of $\mu = 2$.

A display of these 500 chi-square values is now produced by selecting
Graph→Histogram. In the text box under **X**, type *C6*; in **Annotation→Title**, type
Histogram of 500 Chi-square Values with df = 3; in **Annotation→Display Labels** click
on **Display data labels**.

As seen in figure *Minitab 7.8*, min and max of these values are 0.0051 and 12.8617,
respectively. For a histogram with 13 classes, select **Options**, **Cutpoint**, and type
0 : 13 / 1 in the **Define intervals using values** text box.

A display of this histogram is shown in figure *Minitab 7.9*.

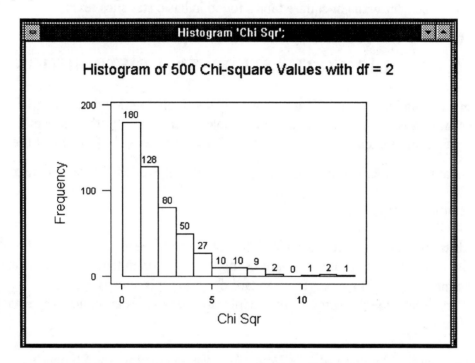

Minitab 7.9: Chi-square Values from the Standard Normal Population

Observe that the shape of this histogram is consistent with the shape of a chi-square
distribution with df = 2 as discussed in most elementary statistics texts.

On can also compare the in the 10th and 90th percentiles of these 500 values with the
theoretical results provided in the chi-square tables. According to these tables with df = 2,
the 10th percentile of the chi-square values is at $\chi^2 = 0.211$ and the 90th percentile is at
$\chi^2 = 4.605$.

To compare this with the 500 Chi-square values obtained in this simulation, sort these
values into column C7 by selecting the commands **Manip→Sort**. Type *C6* in both the
Sort column(s) and **Sort by column(s) in** text boxes. Type *C7* in the **Store sorted
column** text box.

For this particular simulation of the sample of 500 chi-square values, the 50th and 51st values are $\chi^2 = 0.2437$ and $\chi^2 = 0.2502$, respectively, so the 10th percentile of these values is the midpoint 0.2470. The 450th and 451st values are $\chi^2 = 4.4116$ and $\chi^2 = 4.4137$, respectively, so the 90th percentile of these 500 values is the midpoint 4.4127. These empirical percentiles are in fairly close agreement with the theoretical percentiles described above.

The **Lab #7**: *Data Sheet* involves further explorations relating to the distribution of chi-square values for a given degree of freedom. Sampling will be performed from populations which are not necessarily normal. In this event the results obtained may not be in close agreement with chi-square tables found in basic statistics texts.

2. SIMULATING THE STUDENT'S T DISTRIBUTION

When a population is approximately normally distributed with unknown variance and random sampling is performed with a small sample size n (n < 30), the Student's t-statistic, first formulated by W. S. Gosset in 1908, must be used. Recall that the formula for the t-statistic is described by $t = \dfrac{\bar{x} - \mu}{s/\sqrt{n}}$. Tables of Student's t-values are displayed in basic statistics texts.

The distribution of t-values varies with the degrees of freedom, given by df = n-1. These distributions resemble the standard normal distribution in that they are symmetric about the mean $\mu = 0$. However, the standard deviation σ is always greater than 1. As the sample size increases the distribution of t-values approaches the standard normal distribution.

In this section and in the **Lab #7**: *Data Sheet* we will explore the Student's t-distributions through simulation using the Windows version of **MINITAB**.

Example 3: Consider a population which is normally distributed having mean $\mu = 30$ and standard deviation $\sigma = 6$. A simulation will be performed through **MINITAB** of randomly selecting 500 samples of size four from this population in. We will produce columns of sample means, sample standard deviations, and Student's t-values.

The 500 random samples of size four each are obtained by selecting **Calc→Random Data→Normal** and proceeding as in *Example 2*. Enter *C1-C4* in the **Store in column(s)** box, *30* in the **Mean** box, and *6* in the **Standard deviation** box. After clicking on **OK** we will observe four columns in C1, C2, C3, and C4, each having 500 rows of values from the population. Each row of four values can be viewed as a random sample drawn from the population.

Next, place labels of *Mean* for column C5, *St Dev* for column C6, and *t-value* for column C7. Fill the 500 sample means in column C5 by selecting **Calc→Row Statistics**. In the dialog box click on **Mean** and type *C1-C4* in the **Input variables** box. In the **Store results in** box type *C5*. The sample variances are stored in column C6 in the same manner except that we must click on **Standard deviation** and enter *C6* in the **Store results in** box.

The 500 Student's t-values are produced in column C7 by first selecting **Calc→ Mathematical Expressions**. Type *C7* in the **Variable** text box and *(C5 - 30)/(C6/2)* in the **Expression** text box. Note that this expression conforms to the formula for t given by

$$t = \frac{\bar{x} - \mu}{s/\sqrt{n}} \quad \text{for } \mu = 30 \text{ and } n = 4.$$

The mean and standard deviation of these 500 Student's t-values will next be found and compared with what is expected for the population of Student's t-values when the degrees of freedom is three. Moreover, a histogram of these 500 t-values will be generated so that its distribution can be compared with the standard normal distribution.

Select **Stat→Basic Statistics→Descriptive Statistics** and type *C7* in the **Variables** box. Note the values of the mean 0.1302 and standard deviation 1.5512 in the **Session** window as displayed in figure *Minitab 7.10*. These values are reasonable considering that the population mean of the Student's t-distribution is $\mu = 0$ and the population standard deviation σ is greater than 1.

```
┌─────────────────────────────  Session  ──────────────────────────┐
│ MTB > Random 500 C1-C4;                                           │
│ SUBC>   Normal 30.0 6.                                            │
│ MTB > RMean C1-C4 C5.                                             │
│ MTB > RStDev C1-C4 C6.                                            │
│ MTB > Let C7 = (C5-30)/(C6/2)                                     │
│ MTB > Describe  't-value'.                                        │
│                                                                  │
│                 N      MEAN    MEDIAN    TRMEAN     STDEV   SEMEAN │
│ t-value       500    0.1302    0.0968    0.0876    1.5512   0.0694 │
│                                                                  │
│                MIN       MAX        Q1        Q3                   │
│ t-value    -6.2065   13.9710   -0.6863    0.7803                   │
│                                                                  │
│ MTB > |                                                          │
└──────────────────────────────────────────────────────────────────┘
```

Minitab 7.10

A histogram of these 500 Student's t-values is produced through the commands **Graph→Histogram**. From the **Session** window in figure *Minitab 7.10* it is observed that the t-values which are within three standard deviations of its mean is the interval (-4.5234, 4.7838).

For a histogram with 10 classes starting at -5 and ending at 5, click on **Options**, choose **Cutpoint**, and type *-5 : 5 / 1* in the **Define intervals using values** box.

A histogram for this simulation is displayed in figure *Minitab 7.11*. Note the similarities between this histogram and the standard normal distribution shape. This histogram is bell-shaped and nearly symmetric about zero. Also, the t-values have a much greater dispersion that the standard normal values.

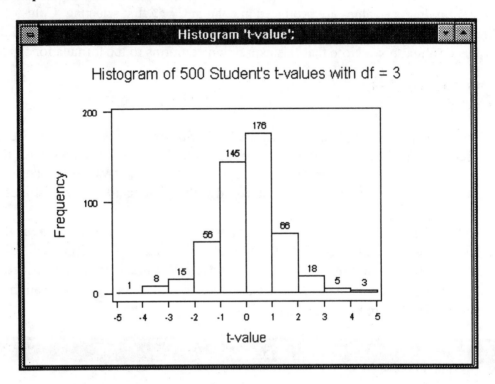

Minitab 7.11: Sampling from a Normal Population with $\mu = 30$, $\sigma = 6$

Recall that approximately 95% of the standard normal values lie in the interval (-2, 2). Observe from the figure *Minitab 7.11* histogram, 50 of the 500 t-values lie outside this interval and thus only 90% of the t-values fall within the interval (-2, 2).

According to a table of Student's t-values with degrees of freedom df = 3, 90% of the t-values are expected to lie in the interval (-2.353, 2.353) which differs somewhat from what was obtained from the above sample. As a reminder, note the following statement.

Warning: There is <u>no guarantee</u> of obtaining a close approximation to a probability based on a population when using an empirical probability from a random sample from the population.

154

LAB #7: *DATA SHEET*

Name: _____   *Date:* _____

Course #: _____   *Instructor:* _____

1. **Team Exploration Project** - An experiment consists of randomly tossing a fair die. Let X be the random variable which records the number of dots which show up. Turn in a detailed report of your results for this project after carrying out the following procedures.

   (a) Fill in the following probability distribution table for X. Using the formulas given in *Lab #4*, find the mean, standard deviation, and variance of X.

x-value						
P(X = x)						

   μ = _____. σ = _____. σ^2 = _____.

   (b) Simulate the die toss experiment 30 times, record the sample standard deviation s and the sample variance s^2. This is to be performed 500 times using **MINITAB**. If only the student version is available, reduce the number of columns or rows to a manageable number. Select **Calc→Random Data→Integer** to generate 30 columns of values in C1 - C30. Enter *1* in the **Minimum value** text box and *6* in the **Maximum value** text box. Each column will have 500 integers values from among the integers 1, 2, 3, 4, 5, 6.

   (c) In the manner discussed earlier in this session, generate the standard deviations of the 500 rows in column C31(label: *St Dev*) and variances in column C32 (label: *Variance*).

   (d) Use **Stat→Basic Statistics→Descriptive Statistics** in order to determine the min, max, mean, and standard deviation of the 500 variances. Provide a printout of the **Session** window and highlight these statistics. Discuss.

(e) Compare the value of the mean of these 500 variances in column C32 with the value of σ^2 found in part (a). Based on this simulation, is it reasonable to conclude that S^2 is an unbiased estimator of σ^2? Discuss.

(f) From the information obtained in part (d), provide a printout of a histogram of the 500 variances in column C32. The histogram should have approximately 15 bars; the **CutPoint** option should be selected.

(g) In general, the sample standard deviation is a biased estimator of the population standard deviation. As mentioned earlier, it becomes closer to becoming unbiased as the sample size increases. Use **Stat→Basic Statistics→Descriptive Statistics** on the sample standard deviations in column C31 to explain how the mean of C31 compares with the value of σ found in part (a). Does the above statement about the bias of S seem to hold true in your simulation? Discuss.

2. **Team Exploration Project**. Carry out the experiment, similar to *Example 3*, of generating 500 samples of size four from the standard normal population. Turn in a detailed report of your results for this project.

(a) Store the random values in columns C1 - C4. Generate the 500 sample means in column C5 (label: *Mean*) and sample standard deviations in column C6 (label: *St Dev*). Select **Calc→Mathematical Expressions** to store the Student's t-values in column C7 (label: *t-value*).

(b) Apply **Stat→Basic Statistics→Descriptive Statistics** on column C7 in order to determine the min, max, mean, and standard deviation of the 500 t-values. Provide a printout of the **Session** window and highlight these statistics. The population mean of a Student's t distribution is always $\mu = 0$. The population standard deviation when $n > 3$ is given by the formula $\sigma = \sqrt{\dfrac{n-1}{n-3}}$. Compare the mean and standard deviation of the 500 t-values with the values of μ and σ.

(c) From the information obtained in part (b), provide a printout of a histogram of the 500 t-values in column C34. The histogram should have approximately 15 bars; the **CutPoint** option should be selected. Compare the shape of the histogram with what you would expect for a Student's t distribution having df = 3.

(d) In the manner of *Example 2*, determine the 10th and 90th percentiles of the 500 t-values. Compare these percentiles with the corresponding percentiles found in the Student's t tables for df = 3. Explain how your results were obtained.

3. **Team Exploration Project**. Suppose that a certain population has an exponential distribution with mean $\mu = 6.5$ (see *Lab #5*). The standard deviation is equal to the mean, so $\sigma = 6.5$ hours. Turn in a detailed report of your results for this project.

(a) Use **MINITAB** to simulate the selection of 500 random samples, each of size 30, from this population by selecting **Calc→Random Data→Exponential**. Store these columns of values in C1-C30. If the student version of **MINITAB** is being used, reduce the number of rows and/or columns accordingly.

(b) Store the column of 500 sample standard deviations in column C31 (label: *St Dev*), the sample variances in column C32 (label: *Variance*), and the sample medians in column C33 (label: *Median*). Do this by selecting **Calc→Row Statistics** and **Calc→Mathematical Expressions**.

(c) Select **Stat→Basic Statistics→Descriptive Statistics** on C31-C33 to generate various statistics relating to the 500 sample standard deviations, variances, and medians. Provide a printout of the **Session** window and highlight the min, max, mean, and standard deviation of each of the three columns.

(d) Compare the mean of the 500 sample standard deviations with the population standard deviation σ. Does it appear that S is an unbiased estimator of σ? Explain. [See Problem 1, part (g).]

(e) Compare the mean of the 500 sample variances with the population variance σ^2. Does it appear that S^2 is an unbiased estimator of σ^2? Explain.

(f) Compare the mean of the 500 sample medians with the population median $\tilde{\mu}$. For an exponential distribution with mean μ, one has that $\tilde{\mu} = \ln(2)*\mu \approx .693\mu$. Does it appear that the sample median $\tilde{x}$ is an unbiased estimator of the population median $\tilde{\mu}$? Explain.

(g) From the information obtained in part (c), provide printouts of histograms of columns C31 - C33 which contain the 500 standard deviations, variances and medians, respectively. The histograms should have approximately 15 bars; the **CutPoint** option should be selected. Explain your procedure. Describe the shapes of each of the histograms.

4. **Team Exploration Project**. Consider a continuous uniform distribution over the interval [10, 30] to sample from (see *Lab #5*). Turn in a detailed report of your observations.

(a) Determine the values of the population mean μ, variance σ^2, and standard deviation σ. Draw a picture of the density function for this distribution.

(b) Use **MINITAB** to store 500 random samples of size 30 in columns C1-C30, selected from this population.

(c) Investigate the chi-square distribution by considering sample sizes 5 (using C1-C5), size 10 (using C1 - C10), and size 30 (using C1 - C30). Store the 500 chi-square values for each of these three sample sizes by using the procedure discussed in *Example 2*.

(d) Generate the mean and standard deviation for each of these three columns of chi-square values. Compare these values with the population mean and standard deviation of the chi-square distribution with df = n - 1 which are given by the formulas $\mu = n - 1$ and $\sigma = \sqrt{2(n-1)}$.

(e) Provide printouts of three histograms of the 500 chi-square values for the three sample sizes n = 5, n = 10, and n = 30. Describe the shape of these distributions.

(f) Look up the 10th and 90th percentile values in the chi-square table for n = 5, n = 10, and n = 30. Find these percentiles for each of the three columns of chi-square values and compare with the table values.

(g) From columns C1-C5, C1-C10, and C1-C30, create three columns of **sample ranges**. This option is available by selecting **Calc→Row Statistics**.

(h) Find the mean of each column of sample ranges and compare them to the population range of this uniform population. Does it appear that sample range is an unbiased estimator of the population range for uniform populations? Explain.

(i) Provide histogram printouts of the three columns of sample ranges. Describe the shape of each histogram.

STATISTICS LAB # 8

HYPOTHESIS TESTS FOR THE POPULATION MEAN - ONE POPULATION

PURPOSE - to use MINITAB to

1. **test** a **population mean** μ when the population **standard deviation** σ is **known**
2. **test** a **population mean** μ when the population **standard deviation** σ is **unknown**

BACKGROUND INFORMATION

1. **Parametric tests** - hypothesis tests for the value of a population parameter.

2. **Null Hypothesis (H_0)** - a statement of a zero or null difference that is directly tested. This will correspond to the original claim if that claim includes the condition of no change ($=$) or no difference such as $\geq$ or $\leq$. We test the null hypothesis directly since the final conclusion will be either rejection of H_0 or failure to reject H_0.

3. **Alternative Hypothesis (H_1)** - a statement that must be true if the null hypothesis is false.

4. **Test Statistic** - a sample statistic or a value based on the sample data. It is used in making the decision about rejecting or failing to reject the null hypothesis.

5. **Decision Rule (Critical region, Rejection region)** - this defines the range of the values of the test statistic that leads to the rejection of the null hypothesis.

6. **Critical Value(s)** - the boundary value(s) that separates the critical region from the values of the test statistic that would lead to failing to reject the null hypothesis.

7. **Type I Error** - rejection of a true null hypothesis. The probability of a **Type I Error** is usually denoted by the symbol α (alpha).

8. **Type II Error** - failing to reject the null hypothesis when it is false. The probability of a **Type II Error** is usually denoted by the symbol β (beta).

9. **Significance Level** - the same as the probability of a **Type I Error**; denoted by α. Typical values of α are 0.01, 0.05 or 0.10.

159

10. **Prob Value (p-value)** - the smallest value of α (probability of a Type I Error) that would result in the rejection of the null hypothesis based on a given test statistic.

11. **P-value Decision Criterion:**

   (i) If p-value $\leq \alpha$, reject the null hypothesis.
   (ii) If p-value $> \alpha$, fail to reject the null hypothesis.

PROCEDURES

First load the **MINITAB** (windows version) software as described in *Lab #0*.

1. TESTING THE POPULATION MEAN μ WHEN THE POPULATION STANDARD DEVIATION σ IS <u>KNOWN</u>

NOTE: **For sample sizes greater than or equal to 30, the 1-Sample Z option *may* be used since the population standard deviation σ can be approximated by the sample standard deviation S. The population standard deviation is available only in rare situations. However, since we have the capability of the MINITAB technology, we should use the 1-Sample t option to test for the mean when σ is unknown even for large sample sizes.**

Example 1: In some societies, a relationship is believed to exist between the length of the *lifeline* on your palm and the age at which you die. A study on this relationship was done by Mather and Wilson (*JAMA*, **229**, No. 11, 1974, 1421 - 1422). *Note:* This data set is given in Appendix A. **MINITAB** is used to test the hypothesis that the **average length** of one's lifeline on the left-hand is *greater than 8.5 centimeters*. Assume that the population *standard deviation is $\sigma = 1.4$ cm* and that the level of significance $\alpha = 0.05$. *The values of 8.5 and 1.4 were obtained from a sample of students (live!) from one of our elementary statistics class. The values of 8.5 and 1.4 are used to help illustrate the procedures.*

Here, we will use the **1-Sample Z** option since just to illustrate the procedure. Enter the data set with the *AGE* values in C1 and the *LENGTH* values in C2. To test whether there is significant evidence to claim that the average length (μ) of the lifeline is indeed greater than 8.5 cm (right-tail test), you need to set up the following null and alternative hypotheses:

$$H_0: \quad \mu = 8.5 \text{ (cm)} \quad (\leq)$$

$$H_1: \quad \mu > 8.5 \text{ (cm)}$$

To use **MINITAB** to perform the hypothesis test, select **Stat→Basic Statistics→ 1-Sample Z**. The **1-Sample Z** dialog box appears. Use the mouse to select C2 *(LENGTH)* to list in the **Variables** text box. Select **Test mean** and type in 8.5 in the **Test mean** text box. From the **Alternative** drop down menu, select **greater than**. In the **Sigma** text box, type in 1.4. The **1-Sample Z** dialog box with the appropriate entries is shown below in figure *Minitab 8.1*.

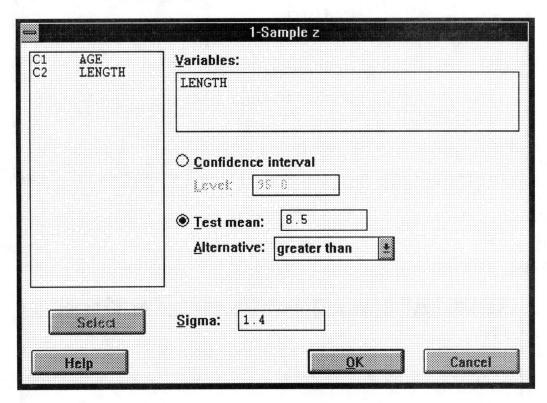

Minitab 8.1

Select the **OK** button. Following the traditional (or classical) procedure found in most elementary statistics text books, you need to determine the **critical z-value**. Select **Calc →Probability Distributions→Normal→Inverse cumulative probability**. Next select the **Input constant** option, type in 0.95 in the text box, and select the **OK** button. This is equivalent to finding the z-score (z_0) such that $P(z \leq z_0) = 0.95$. The **Normal Distribution** dialog box with the appropriate entries is shown in figure *Minitab 8.2*.

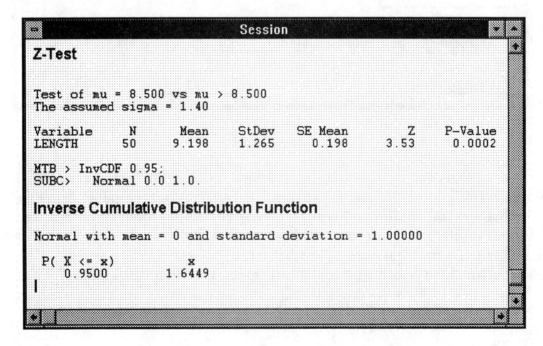

Minitab 8.2

The **Session** window below in figure *Minitab 8.3* displays the **MINITAB** commands and output corresponding to the previous menu choices.

Z-Test

```
Test of mu = 8.500 vs mu > 8.500
The assumed sigma = 1.40

Variable      N      Mean     StDev    SE Mean        Z     P-Value
LENGTH       50     9.198     1.265      0.198     3.53      0.0002

MTB > InvCDF 0.95;
SUBC>    Normal 0.0 1.0.
```

Inverse Cumulative Distribution Function

```
Normal with mean = 0 and standard deviation = 1.00000

  P( X <= x)          x
     0.9500       1.6449
```

Minitab 8.3

The subcommand **SUBC> Alternative 1** indicates that the alternative hypothesis is **right-tail**. Also, a **-1** or **0** will represent a **left-tail** or a **two-tail** test, respectively. If the null hypothesis is true, the sample mean of 9.198 has a z-score of 3.53. The critical z-value is 1.6449, so we **reject** H_0 since the test statistic value of 3.53 is greater than 1.6449. We conclude that the **average length of the lifeline on the left hand is indeed greater than 8.5 cm at the 0.05 level of significance.**

If we use the **p-value** criterion, observe that the **p-value** = 0.0002 < 0.05, so we have the same conclusion as above.

Exercise 1: *(See Problem 1).* An unopened sample of 1989-1990 Fleer basketball cards was obtained by purchasing four packs of these cards. After duplicates were removed, a sample of 45 cards remained. The value of each card was determined from the Beckett Basketball Monthly (April 1991). This publication was printed at the completion of the 1990-1991 National Basketball Association season. The information collected from these cards are given in Appendix B. The variables are as follows:

> PRICE - value of the card
> NYEARS - the number of years the player had played pro basketball at
>   the time the cards were printed
> HEIGHT - the height of the player in feet
> WEIGHT - the weight of the player rounded to the nearest 5 pounds
> APSPG - the player's career average points scored per game
> ARPG - the player's career average rebounds per game

Use **MINITAB** to test, at the 0.02 level of significance, whether the average height of basketball players is greater than 6.5 feet when $\sigma = 0.3$. The value of 6.5 was chosen since most NBA basketball players have a height between 6 feet and 7 feet.

2. TESTING THE POPULATION MEAN μ WHEN THE POPULATION STANDARD DEVIATION σ IS <u>UNKNOWN</u>

This section deals with hypothesis tests for the population mean when the population standard deviation is unknown.

Example 2: Based on final averages for an elementary *Business Statistics* course over several years, Dr. J, who taught the course found that the average score was 73. He claims that his last class was a below-average class. Test his claim, at the 0.05 level of significance. The averages for his last class are given below. These averages were rounded up to the next integer and we will assume that this sample was drawn from an approximately normal distribution.

Student	1	2	3	4	5	6	7	8	9	10	11	12	13	14	15	16	17
Average	51	73	85	87	63	50	73	84	62	51	79	75	63	85	79	81	83

Since you are required to determine whether this set of average scores reflect a below-average class, you need to test whether there is significant evidence to claim that the average exam scores is indeed less than 73 (left-tail test). You need to set up the following null and alternative hypotheses:

$$H_0: \mu = 73 \ (\geq)$$

$$H_1: \mu < 73$$

You need to perform a **1-Sample t** test since the **variance of the population is unknown**. To use **MINITAB** to perform the hypothesis test, select **Stat→Basic Statistics→1-Sample t**. The **1-Sample t** dialog box will appear. Enter the averages in C1 *(EXAMAVG)*. Select **Test mean** and type *73* in the **Test mean** text box. From the **Alternative** drop down menu, select **less than**. The **1-Sample t** dialog box with the appropriate entries is shown below in figure *Minitab 8.4*.

Minitab 8.4

Select the **OK** button. Following the traditional (or classical) procedure found in most elementary statistics text books, you need to determine the **critical t-value**. Select **Calc→Probability Distributions→T→Inverse cumulative probability**. In the

Degrees of freedom text box type *16* (since df = *n* - 1 = 17 - 1 = 16). Next select the **Input constant** option, type *0.05* in the text box, and select the **OK** button. The **Session** window below in figure *Minitab 8.5* displays the **MINITAB** commands and output corresponding to the previous menu choices.

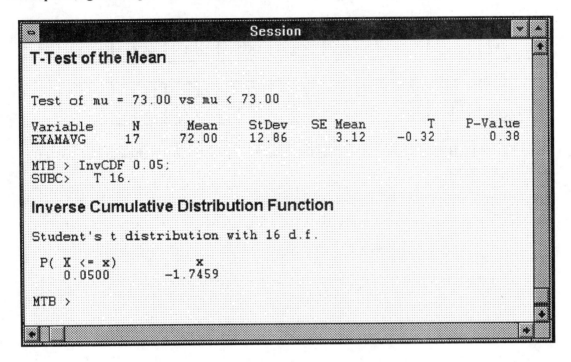

Minitab 8.5

The subcommand **SUBC> Alternative -1** indicates that the alternative hypothesis is **left-tail**. If the null hypothesis is true, the sample mean of 72 has a t-score of -0.32. The critical t-value is -1.7459, so we **fail to reject** H_0 since the test statistic value of -0.32 is greater than -1.7459. We can conclude that there is **insufficient evidence** to infer that the **average test scores is less than 73**. Thus, there is insufficient evidence to claim that this class is operating at a level that is below average.

If we use the **p-value** criterion, observe that the **p-value** = 0.38 > 0.05, so we have the same conclusion as above.

Exercise 2: *(See Problem 2).* The quality-control department of a canning company requires that the mean weight of a specific canned product be 16 ounces. Fifteen cans filled by the same machine were sampled at random from the assembly line and their weights measured. The results are given below:

Can	1	2	3	4	5	6	7	8
Weight	16.5	16.2	16.9	15.8	16.2	16.4	16.5	16.8

Can	9	10	11	12	13	14	15	
Weight	15.8	15.5	16.7	15.5	16.2	16.7	15.7	

Use **MINITAB** to test the claim that the machine is *over-filling* the cans at the 0.02 level of significance.

LAB #8: *DATA SHEET*

Name: _____   *Date:* _____

Course #: _____   *Instructor:* _____

1. Write up the hypothesis test for *Exercise 1* using the *critical region* approach. State your conclusion in the context of this problem.

    H_0: _____ vs. H_1: _____

    Test Statistic: _____

    Decision Rule:

    Conclusion:

2. Write up the hypothesis test for *Exercise 2* using the *p-value* approach. State your conclusion in the context of this problem.

    H_0:

    H_1:

    Test Statistic: _____

    Decision Rule:

Conclusion:

3. One of the basic assumptions for these hypotheses tests is that the sample is drawn from a normal or approximately normal distribution. Discuss whether the normality assumption holds for the samples in *Exercises 1* and *2*. Discuss the NPPs (see *Lab #6*) and also observations obtained from the histograms for these samples. Provide hard copies of the graphs with your discussion.

4. In the February 21, 1994 issue of Barron's Financial Report, a systematic sample of 15 stocks from NASDAQ provided the following information about their price per share and volume (vol., in 100s) for the week:

Stock #	1	2	3	4	5	6	7	8
Price	9 7/8	20 1/2	24 3/8	25 5/8	20 1/4	34 3/4	20	12 3/8
Vol.	1,574	10,574	401	77	277	82	197	6,205

Stock #	9	10	11	12	13	14	15	
Price	27	4 3/4	7	2 3/4	4 1/4	9	48 3/4	
Vol.	6,205	622	842	1,348	5,957	153	4,816	

Enter this data in two columns named *PRICE*, and *VOLUME*. In each of the following, set up the hypothesis test and arrive at a conclusion in two ways: (1) by using the p-value; (2) by using the critical region (classical). State your conclusion in the context of this problem.

(a) Use the PRICE data to test the hypothesis, at the 0.05 level of significance, that the mean price per share of the NASDAQ market was greater than $15 per share.

(1) *p-value approach*

H$_0$:

H$_1$:

Test Statistic: _____

Decision Rule:

Conclusion:

(2) *critical region (classical) approach*

H₀:

H₁:

Test Statistic: _____

Decision Rule:

Conclusion:

(b) Use the **VOLUME** data to test the hypothesis, at the 0.1 level of significance, that the mean NASDAQ volume (in hundreds of shares) is less than 4,000.

(1) *p-value approach*

H₀:

H₁:

Test Statistic: _____

Decision Rule:

Conclusion:

(2) *critical region (classical) approach*

H₀:

H₁:

Test Statistic: _____

Decision Rule:

Conclusion:

5. This exercise will use a simulation technique to do a hypothesis test. Assume that we have a sample of 20 lifeline lengths of the left hand with a sample mean of 9 cm. The population of all lifelines has a standard deviation of 1.4 cm. At the 0.05 significance level, you are to test the claim that the sample comes from a population with a mean greater than 8.5 cm. Instead of the classical (critical region) or p-value approach, use **MINITAB** to determine whether a sample mean of 9 cm. could easily occur by chance.

Follow the following steps to establish a conclusion:

Step 1: Use **MINITAB** to generate 25 samples of size 20 from a normal distribution with a mean of 8.5 and a standard deviation of 1.4.

Step 2: Compute the sample means for each of these 25 samples.

Step 3: Examine the 25 sample means and determine whether or not a sample mean of 9 can *easily* occur by chance or whether it is significantly different from 8.5
Note: "Easily" here means whether you observe a sample mean of 9 or above more than 5% of the time.

Step 4: Repeat the *Steps 1 - 3* five times.

Step 5: Formulate a conclusion.

Present your results in a report.

STATISTICS LAB # 9

CONFIDENCE INTERVALS FOR THE MEAN - ONE POPULATION

PURPOSE - to use MINITAB to

1. construct a **confidence interval** for a population mean
2. enhance student understanding of a **point estimate** of a population mean by random sampling from a variety of populations
3. simulate the selection of a large number of confidence intervals in exploring **level of confidence**
4. select a random sample from a Bernoulli population to generate a confidence interval for a population proportion.

BACKGROUND INFORMATION

1. **Point estimate of a parameter -** a sample statistic obtained from a random sample of a population. The sample statistic provides an estimate of a population parameter. In particular, the sample mean $\overline{x}$ is a **point estimate** of the population mean μ and the sample proportion $\hat{p}$ is a **point estimate** of population proportion p.

2. **Confidence interval for a parameter -** an interval of values containing the point estimate of a parameter.

3. **Level of confidence for a confidence interval -** the percentage $(1 - \alpha)*100\%$ of the time that the parameter should fall within a generated set of confidence intervals for this parameter. The most common confidence levels are:

    90% ($\alpha = 0.1$), 95% ($\alpha = 0.05$), and 99% ($\alpha = 0.01$)

4. **Maximum error of the estimate -** the value E such that a confidence interval for μ is expressible as $(\overline{x} - E, \overline{x} + E)$, where $\overline{x}$ is a point estimate of μ. For a proportion p the confidence interval is $(\hat{p} - E, \hat{p} + E)$ where $\hat{p}$ is a point estimate of p.

5. **A critical value from the standard normal distribution -** a value z_α such that the proportion of z-values greater than z_α is equal to α.

6. **A critical value from the Student's t distribution with df = n - 1 degrees of freedom** - a value t_α such that the proportion of t-values greater than t_α is equal to α.

7. **Formulas for the maximum error E -**

(a) $E = z_{\alpha/2}(\sigma/\sqrt{n})$ (σ is known; normally distributed sample means)

(b) $E = t_{\alpha/2}(s/\sqrt{n})$ (σ is unknown; approximately normally distributed population if the sample size is small)

(c) $E = z_{\alpha/2}\sqrt{\dfrac{p(1-p)}{n}}$ (large samples; in practice, the proportion p will be replaced by the point estimate $\hat{p}$ in the formula)

PROCEDURES

MINITAB will be used to generate confidence intervals for a population mean and population proportion. To access **MINITAB**, refer to *Lab #0*.

1. CONFIDENCE INTERVALS FOR A POPULATION MEAN WITH KNOWN POPULATION STANDARD DEVIATION σ

When the population standard deviation σ is known and the distribution of the sample means for a fixed sample size are normal or approximately normal, the **MINITAB** commands **Stat→Basic Statistics→1-Sample Z** can be used to generate confidence intervals for μ. The sample of data values must first be entered in one of the columns in the **Data** window.

Example 1: A population is normally distributed with a known standard deviation σ = 16.0. A random sample of size 20 yields the following data values:

55	43	69	32	77	58	84	80	75	61	88	93	77	68	49	72	59	81	79	62

From the conditions imposed, the distribution of sample means is normally distributed and σ is known. In order to generate a 95% confidence interval for μ, enter the data values in column C1 (label: *Data*), select the **MINITAB** menu options **Stat→Basic Statistics→1-Sample Z**, and fill in the text boxes as displayed in figure *Minitab 9.1*.

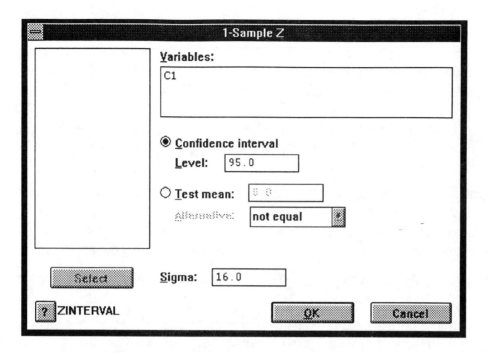

Minitab 9.1

Since this **1-Sample Z** dialog box generates both confidence intervals and hypothesis test information for a population mean, be sure to click on the **Confidence Interval** button and type in the desired level of confidence. If a different level of confidence in desired, type this percentage in the **Confidence Interval** box. After the **OK** button is clicked, the **Session** window will provide us with the information shown in figure *Minitab 9.2*.

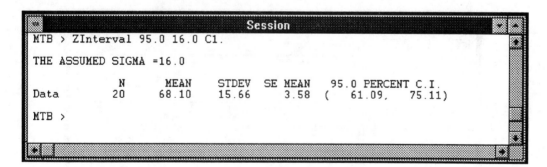

Minitab 9.2

As a consequence of these results, one can be 95% confident that the population mean μ falls inside the interval (61.09, 75.11).

Observe the other information provided in the **Session** window, namely the number of data values N = 20, the sample mean **MEAN** = 68.10, the standard deviation **STDEV** = 15.66 and the standard error of the mean **SE MEAN** = $\sigma / \sqrt{n}$ = 3.58.

By using the formula for the maximum error of the estimate given by $E = z_{\alpha/2}(\sigma/\sqrt{n})$ with $z_{\alpha/2} = z_{.025} = 1.96$, the value of E is 7.01. One then verifies the correctness of the above confidence formula using $\bar{x} = 68.10$ in $(\bar{x} - E, \ \bar{x} + E)$.

2. CONFIDENCE INTERVALS FOR A POPULATION MEAN WITH UNKNOWN σ

In practice, the value of the population standard deviation σ will not be known. In the event that the sample size is small ($n < 30$), the Student's t-statistic with df = n-1 degrees of freedom should be used to generate a confidence interval for the population mean μ. Recall that once a random sample of size n has been chosen from the population, the value

of t is given by the formula $\quad t = \dfrac{\bar{x} - \mu}{s/\sqrt{n}}$. To use the t-statistic one must assume that

the population is normal or approximately normal. This can be tested by generating a histogram from the sample data. A more reliable approach is to produce a normal plot of the sample data by selecting **Graph→Normal Plot** as indicated in figures *Minitab 6.9* and *Minitab 6.10*. The procedure for carrying this out was discussed in **Lab #6** and will be addressed for the data in *Example 1*.

Example 1 (Continued): Assume that the value of the population standard deviation is unknown and that a 95% confidence interval for μ is to be generated. By selecting **Stat→Basic Statistics→1-Sample t**, a dialog box will appear as displayed in figure *Minitab 9.3*. Fill in the text boxes as indicated.

Minitab 9.3

After clicking the **OK** button, the results will be displayed in the **Session** window as shown in figure *Minitab 9.4*.

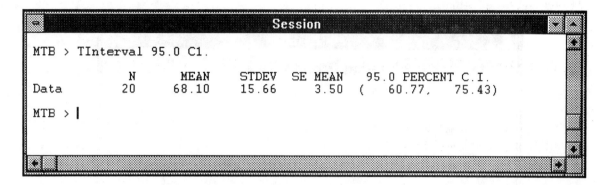

```
                                  Session
MTB > TInterval 95.0 C1.

               N      MEAN      STDEV   SE MEAN     95.0 PERCENT C.I.
Data          20      68.10     15.66    3.50    (   60.77,    75.43)

MTB > |
```

Minitab 9.4

Based on this information, one can be 95% confident that the population mean μ lies between 60.77 and 75.43.

Observe that this confidence interval (60.77, 75.43) is somewhat larger than the one generated for known σ. This is a consequence of the fact that the maximum error formula $E = t_{\alpha/2}(s/\sqrt{n})$ uses the larger t-critical value $t_{\alpha/2}$ as opposed to the z-critical value $z_{\alpha/2}$ which is used for known population variance or large sample sizes.

As mentioned earlier, the population should be normal or approximately normal in order to obtain reliable results when generating a confidence interval of the mean using the Student's t distribution. We will use the **Graph→Normal Plot** options in order to see if the 20 data values in *Example 1* appear normally distributed. Fill in the boxes in the **Normal Probability Plot** window as indicated in figure *Minitab 9.5*.

Minitab 9.5

177

After clicking the **OK** button a probability plot of these 20 data values will appear in the **Normplot** window as displayed in figure *Minitab 9.6* when selecting the Anderson-Darling test for normality.

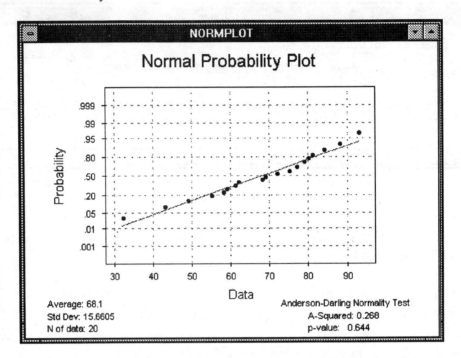

Minitab 9.6

As mentioned in *Lab #6*, when the data appears to cluster about the plotted line in the graph, the population should have an approximately normal distribution. More precisely, set up the hypothesis test

H_0: "the population is normal"   vs. H_1: "the population is not normal".

The p-value of 0.664 indicates that H_0 should not be rejected at any of the standard levels of significance. This demonstrates that we are justified in assuming that the population in this exercise is approximately normally distributed.

3. GENERATING CONFIDENCE INTERVALS BY RANDOM SAMPLING

Populations such as the normal, uniform, exponential, and discrete can be sampled using the **MINITAB** commands **Stat→Random Data**. The sample mean $\bar{x}$ of the data generated will then provide a **point estimate** for μ. If this process is applied repeatedly, a large number of confidence intervals can be produced in order to determine the percentage of them which contain the population mean μ. This percentage will be compared with the specified level of confidence.

Example 2: Generate 200 random samples, each of size 10, from a normal population whose mean and standard deviation are $\mu = 50$ and $\sigma = 15$, respectively. From each of the 200 sample means, a 90% confidence interval for μ will be found using the known value of σ.

Select the commands **Calc→Random Data→Normal** in order to generate 200 rows and 10 columns of data in C1-C10 as indicated in figure *Minitab 9.7*.

Normal Distribution	
Generate	200	**rows of data**
Store in column(s):	
C1–C10	
Mean:	50.0
Standard deviation:	15.0

Minitab 9.7.

In order to produce the 200 sample means in column C11 (label: *Mean*), select **Calc→Row Statistics,** click on the **Mean** button, type *C1-C10* in the **Input variables** text box, and type *C11* in the **Store result in** text box.

Select **Calc→Mathematical Expressions** in order to store the value of the maximum error of the estimate $E = z_{\alpha/2}(\sigma/\sqrt{n})$ in constant K1. Since 90% confidence intervals are being constructed with n = 10, $\sigma = 10$, and $z_{.05} = 1.645$, the maximum error is given as $E = (1.645)(15/\sqrt{10})$.

In the **Mathematical Expressions** dialog box, type *K1* for **Variable** and *1.645 * 15 / sqrt(10)* for **Expression**. To see the value of the maximum error K1 in the **Session** window, select **File→Display Data** and type *K1* in the variables/constants box. After selecting **File→Display Data** and entering *K1*, figure *Minitab 9.8* reveals that K1 = 7.80292.

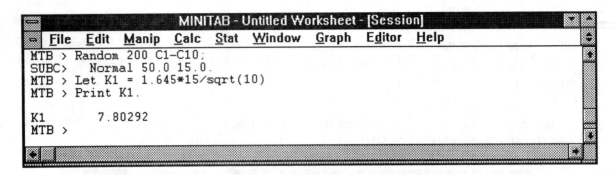

Minitab 9.8

The lower and upper bounds of the 200 confidence intervals obtained from the sample means in column C11 will now be produced and stored in columns C12 (label: *Lower*) and C13 (label: *Upper*), respectively.

To obtain the 200 lower bounds, select **Calc→Mathematical Expressions,** type *C12* in the **Variable** text box and *C11 - K1* in the **Expression** text box. For the upper bounds, type *C13* in the **Variable** text box and *C11 + K1* in the **Expression** text box.

In order to efficiently count the number of confidence intervals which do not contain the mean $\mu = 15$, we will sort columns C12 and C13 and store the results in columns C14 (label: *SortL*) and C15 (label: *SortU*), respectively. The sorting procedure is displayed in figure *Minitab 9.10*. After selecting **Manip→ Sort**, fill in the **Sort** dialog boxes as indicated.

Minitab 9.10

By scanning through columns C14 and C15, the beginning confidence intervals will lie to the left of μ and the end confidence intervals will lie to the right of μ. In the current simulation, a total of 22 of the 200 confidence intervals (11%) did not contain the population mean $\mu = 50$. Since the level of confidence is 90% and 89% of the 200 generated contained μ, this is close to what one would expect.

Note: If this simulation is repeated, different results will be obtained!

We will next display a graph of the boundaries of these 200 confidence intervals by selecting **Graph→Plot**. Also displayed is the horizontal line at $\mu = 50$. This display will allow us to visualize each confidence interval as a vertical line segment. Moreover, we will be able to note the number of times $\mu = 50$ falls within a given confidence interval.

First, label column C16 as *Pop Mean*, select **Calc→Set Patterned Data**, type *C16* in the **Store result in column** text box, and click on the **Arbitrary list of constants** button. Type *50* in the **Arbitrary list of constants** text box, *1* in the **Repeat each value** text box, and *200* in the **Repeat the whole list** text box. Next, label column C17 as *CI No.* and enter the numbers 1, 2, 3, ..., 200 in this column by selecting **Calc→Set Patterned Data**. Click on the **Patterned sequence** button, type in *1* and *200* in the **Start at** and **End at** text boxes, and type *1* in the **Repeat each value** and **Repeat the whole list** text boxes. Note that column C16 contains the value $\mu = 50$ in each of its 200 rows.

A display of the lower and upper bounds of the 200 confidence intervals is found by selecting **Graph→Plot** and filling the **Plot** dialog boxes as shown in figure ***Minitab 9.10***.

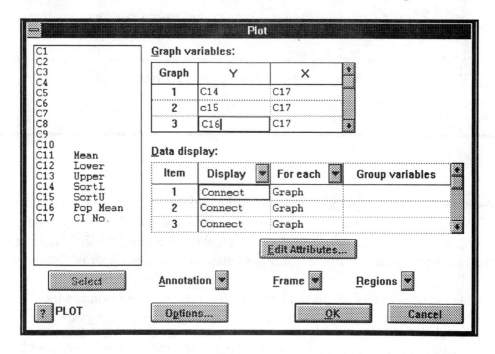

Minitab 9.10

In the **Plot** dialog box in figure *Minitab 9.10* one can enter a title, axis labels, and appropriate tick marks on the axes by selecting **Annotation→Title** , **Frame→Axis →Label**, and **Frame→Tick**. Since graphs of the 200 lower and upper confidence interval boundaries, together with the horizontal line $\mu = 50$ are to be plotted, we will select **Frame→Multiple Graphs** and click on **Overlay graphs on the same page**. Select **Edit Attributes** to declare the type of line to display.

A graph of the lower and upper boundaries of these 200 confidence intervals and the horizontal line at $\mu = 50$ is displayed in figure *Minitab 9.11*.

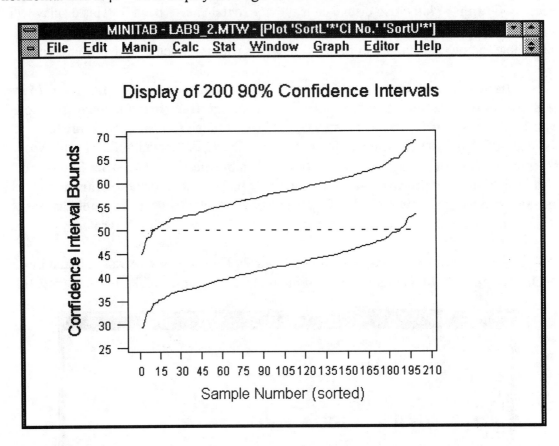

Minitab 9.11

In order to interpret this graph, observe that the horizontal dotted line represents the value of the population mean $\mu = 50$. For each sample number 1, 2, 3, ... 200 indicated on the horizontal axis, there corresponds a 90% confidence interval which is represented by the vertical line segment from the lower boundary curve to the upper boundary curve.

A segment which crosses the dotted line indicates that the mean is inside the confidence interval; otherwise, the mean lies outside the confidence interval.

As seen from figure *Minitab 9.11*, approximately the first 12 sorted confidence intervals and the last eight sorted confidence intervals do not contain the mean for a total of 20. We observed earlier that the actual number of these 200 confidence intervals which do not contain the population mean $\mu = 50$ is 22, or 11% of the total, so that 89% do contain the mean. This is very close to the given 90% level of confidence.

Simulations using **MINITAB** can also produce confidence intervals for a population mean where the Student's t-statistic is used. Random samples are produced in the same manner as discussed in *Example 2*. After the sample means have been produced and stored in one column and the sample standard deviations s in another column, the maximum errors E are generated from the formula $E = t_{\alpha/2}(s/\sqrt{n})$. This is addressed in **Lab #9:** *Data Sheet*.

Confidence intervals for a population proportion p can be generated by **MINITAB** by making random selections with the commands **Calc→Random Data→Bernoulli**.

Example 3: A recent survey indicates that 65% of the adult population are register to vote. The population can be viewed as Bernoulli with $p = 0.65$. The value of p represents the proportion of registered voters; the number 1 corresponds to a registered adult and the number 0 corresponds to a non-registered adult.

MINITAB will now be used to generate a random sample of size 2000 from this population by selecting **Calc→Random Data→Bernoulli**. Figure *Minitab 9.9* shows how to fill in the text boxes in order to produce this sample in column C1 (label: *Sample*).

Minitab 9.12

A confidence interval for population proportion p of successes has the form $(\hat{p} - E, \hat{p} + E)$ where $\hat{p}$ is the sample proportion of successes and E is the maximum error of the estimate $E \approx z_{\alpha/2} \sqrt{\dfrac{\hat{p}(1-\hat{p})}{n}}$. The value of $\hat{p}$ for n trials is given by

$\hat{p} = \dfrac{x}{n}$ where x is the number of successes.

In this experiment success means that the adult is registered to vote and is represented by 1 whereas a failure (not registered) is designated by a 0. The value of x is found by adding the values in column C1. To accomplish this, select **Calc→Column Statistics** and click on the **Sum** button. Enter C1 in the **Input variable** box and type *K1* in the **Store result in** box. After clicking the **OK** button, select **File→Display Data** and type *K1* in the variable/constant box. In the current simulation of this experiment, the **Session** window reveals that K1 = 1295. From this we conclude that the sample proportion of registered voters is $\hat{p} = 0.6475$.

Since a 99% confidence interval for p is desired, the value of $z_{\alpha/2}$ is 2.576. For the maximum error of the estimate E, select **Calc→Mathematical Expressions**, type *K2* in the **Variable** box, and type *2.576*sqrt(.65*(1-.65)/2000)* in the **Expressions** box. The value of E is K2 = 0.02747 in the **Session** window as displayed in figure *Minitab 9.13*.

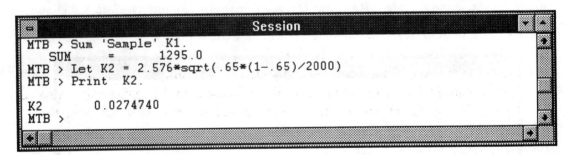

```
                          Session
MTB > Sum 'Sample' K1.
    SUM       =       1295.0
MTB > Let K2 = 2.576*sqrt(.65*(1-.65)/2000)
MTB > Print  K2.

K2        0.0274740
MTB >
```

Minitab 9.13

The 99% confidence interval for the proportion of adults who are registered voters for this simulation is given approximately by (0.6200, 0.6750) as seen from

$$\hat{p} - E, = 0.6475 - 0.0275 = 0.6200 \quad \text{to} \quad \hat{p} + E = 0.6475 + 0.0275 = 0.6750.$$

Note: It must again be emphasized that when this simulation is repeated the results will be different!

Observe that the population proportion p of adults who are registered voters was 0.65 which falls within the above confidence interval. If this simulation was to be performed 100 times, one could expect p = 0.65 to fall within 99 of the 99% confidence intervals generated.

LAB #9: *DATA SHEET*

Name: _____   *Date:* _____

Course #: _____   *Instructor:* _____

1. Load the **MINITAB** data file called *lakes.mtw* into the **Data** window. To accomplish this, select **File→Open Worksheet→Select Worksheet** as discussed in *Lab #0.* In column C2 is located a sample of 71 depths of the lakes surveyed.

  (a) Construct a 90%, a 95%, and a 99% confidence interval for the population mean depth μ of all lakes in the region, using **Stat→Basic Statistics→1-Sample t**. Provide a printout of the **Session** window information relating to these confidence intervals. Compare the sizes of these three confidence intervals and explain why they differ.

  (b) Assume the population standard deviation of the lake depths is 18 feet. Construct a 90%, a 95%, and a 99% confidence interval for the population mean depth of all lakes in the region, using **Stat→Basic Statistics→1-Sample Z**. Provide a printout of the **Session** window information relating to these confidence intervals. Also, use the sample statistic information to generate these confidence intervals using the formula for the standard error $E = z_{\alpha/2} (\sigma /\sqrt{n})$. Compare with the confidence intervals produced by **MINITAB**.

  (c) Explain how and why the 99% confidence interval in part (a) compares to the 99% confidence interval in part (b).

2. Assume that the scores on a standardized exam are normally distributed. A random sample of 26 scores of recent test-takers is described as follows:

15	28	17	30	22	19	21	20	33	14	18	24	26
17	20	35	14	23	19	10	24	29	8	16	18	18

(a) Enter these scores in column C1 (label: *Scores*) in the **Data** window. Use **MINITAB** to produce a 95% confidence interval. Provide a printout of the **Session** window and fill in the blanks below.

Confidence interval: _____.   Sample mean: _____.

Sample standard deviation: _____.   Standard error: _____.

(b) Use the values from part (a) to find the maximum error of the estimate E and use it to generate the confidence interval. How does this compare with the confidence interval results from part (a)?

Maximum error of the estimate: _____.

Confidence interval: _____.

(c) Use **MINITAB** to construct a histogram of the 26 scores using between six and eight classes. Also produce a display of an Anderson-Darling Normal Plot of these scores by following the procedure in figures *Minitab 9.5* and *Minitab 9.6*. Based on this information, is it reasonable to assume that the population of scores is approximately normally distributed? Provide detailed explanations.

3. Consider the normally distribution population from *Example 2*, having mean $\mu = 50$ and standard deviation $\sigma = 10$. Carry out a simulation of 500 99% confidence intervals by selecting samples of size five with **MINITAB**. Provide detailed explanations in a report. If you are using the student edition you may need to reduce the number of simulations to fewer than 500.

(a) Generate 500 random samples, each of size $n = 5$ from this population. Store the data in columns C1 - C5.

(b) Store the sample means of the 500 rows from columns C1-C5 in column C6 (label: *Mean*). Also store the sample standard deviations of the 500 rows from columns C1-C5 in column C7 (label: *St Dev*). It will be necessary for you to choose **Calc→Row Statistics**.

(c) Find the value of $t_{\alpha/2}$. The maximum error of the estimate for the 500 rows of data can be expressed as $E = t_{\alpha/2} * C7/\sqrt{n}$. Select **Calc→Mathematical Expressions** in order to produce a column of E-values in column C8 (label: *E-value*).

$t_{\alpha/2} = $ _____.

Expression for E entered in the dialog box: _____.

(d) Type the expression $C6 - C8$ in the **Mathematical Expressions** dialog box to generate the lower bounds of the 500 confidence intervals into column C9 (label: *Lower*). Also, type the expression $C6 + C8$ in the **Mathematical Expressions** dialog box to generate the upper bounds of the 500 confidence intervals into column C10 (label: *Upper*).

(e) Select **Manip→Sort** to sort the 500 lower bounds and upper bounds of the confidence intervals from columns C9 and C10, respectively, in the manner described in figure *Minitab 9.10*. These sorted bounds should be stored in columns C11 (label: *SortL*) and column C12 (label: *SortU*).

(f) From the information provided in columns C11 and C12, determine the number and percentage of the confidence intervals which do not contain the population mean $\mu = 50$. Is this what you would expect? Explain.

Number for which $\mu = 50$ is to the left of the confidence interval _____.

Number for which $\mu = 50$ is to the right of the confidence interval _____.

Number and percentage of the 500 confidence intervals which do not contain $\mu = 50$

_____.

4. Repeat Problem 3 for an exponential population having mean and standard deviation given by $\mu = \sigma = 10$. Are the results of your simulation consistent with what is expected for 99% confidence intervals? What must be assumed about the population in order to generate confidence intervals involving small samples. Provide a detailed report.

5. Consider the Bernoulli population in *Example 3* where success means the adult selected is registered to vote and the proportion of successes is $p = 0.65$.

(a) Carry out the same simulation using **MINITAB** of randomly selecting 2000 people from this population. Using the same methods as discussed in *Example 3*, generate a 99% confidence interval for p. Provide a printout of the **Session** window. Write down the values of the sample proportion $\hat{p}$, the maximum error of the estimate E, and the 99% confidence interval obtained in your simulation. Explain whether or not the population proportion $p = 0.65$ is inside your confidence interval. Is this what you would expect?

Value of $\hat{p}$ = _____ . Value of E = _____ .

99% confidence interval: _____ .

Interpret this interval:

(b) You will be using a different procedure to generate a 99% confidence interval from the 2000 zero and one values obtained in part (a). The population proportion p is identical to the population μ which means that the commands **Stat→ Basic Statistics→1-Sample t** can be applied to this sample in order to generate a 99% confidence interval. After carrying this out provide a printout of the **Session** window. Write down the values of the sample mean $\bar{x}$, maximum error E, and the 99% confidence interval. Compare $\bar{x}$ with $\hat{p}$, the value of E, and the confidence interval with that obtained in part (a).

Value of $\hat{p}$ = _____ . Value of E = _____ .

99% confidence interval: _____ .

6. Again consider the Bernoulli population described in *Example 3*. The proportion of adults who are registered voters is $p = 0.65$. Perform the following simulation of producing 100 confidence intervals, each having a 90% level of confidence when generating random samples of size 50 from this population.   If you are using the student edition a smaller sample size may be needed. Provide a detailed report of your results.

(a) Select **Calc→Random Data→Bernoulli** to generate 100 random samples, each of size 50 from this population of zeros and ones. Store the data in columns C1-C50.

(b) Generate 100 sample proportions (means) in column C51 (label: *Mean*)

(c) Use the formula $E \approx z_{\alpha/2} \sqrt{\dfrac{\hat{p}(1-\hat{p})}{n}}$ to store the 100 maximum errors of the estimates in column C52 (label: *E-values*). Do this by selecting **Calc→Mathematical Expressions** and enter an expression based on the equation $E = z_{\alpha/2} * sqrt((C51*(1-C51) / n)$. Give the values of n, $z_{\alpha/2}$, used in this formula.

$n =$ _____.

$z_{\alpha/2} =$ _____.

(d) Generate the lower bounds and upper bounds of the 100 90% confidence intervals for proportion in columns C53 (label: *Lower*) and C54 (label: *Upper*), respectively. Follow the procedure described in part (d) of Problem 3.

(e) Sort columns C53 and C54 into C55 and C56 in the manner described in figure *Minitab 9.10* . Label C55 as *SortL* and C56 as *SortU*. From these two columns, determine the percentage of these confidence intervals for which the population proportion p = 0.65 falls inside the interval. Explain whether or not this percentage is consistent with what you would expect based on the given level of confidence in this exercise.

Number with p = 0.65 to the left of the interval _____ .

Number with p = 0.65 to the right of the interval _____ .

Percentage of the 100 confident intervals for which p falls inside _____ .

(f) Provide a printout of a graph of the bounds of the 100 confidence intervals by using the procedure leading up to figure *Minitab 9.13*. The graph should have appropriate title and axis labels. Use the graph to estimate the percentage of these confidence intervals for which p = 0.65 falls inside the interval. Give reasons for your answer.

7. **Team Exploration Project**. Design an experiment based on a population related to the students on your campus by gathering a data sample of size 100. Examples include distance from home to campus, student age, and student GPA. State how your experiment was designed resulting in the selection of the sample from the 100 students polled. Then construct 90%, 95%, and 99% confidence intervals for the population mean by selecting **Stat→ Basic Statistics →1-Sample t** in **MINITAB**. Produce a histogram and a normal plot and use them to explain whether it is reasonable to assume that your population is approximately normally distributed. Provide printouts of the **Session** window and generated graphing window displays. Turn in a detailed report of your results.

STATISTICS LAB # 10

HYPOTHESIS TESTS AND CONFIDENCE INTERVALS FOR TWO POPULATION MEANS

PURPOSE - to use MINITAB to

1. **test** a **difference** of **population means** $\mu_1 - \mu_2$ using **independent samples**
2. **test** a **difference** of **population means** $\mu_1 - \mu_2$ using **dependent samples**
3. generate a **confidence interval** for $\mu_1 - \mu_2$ from **independent samples**
4. generate a confidence interval for $\mu_1 - \mu_2$ from **dependent samples**

BACKGROUND INFORMATION

1. **Independent samples** - the random samples selected from the two populations are selected independently from one another.

2. **Dependent samples** - the random samples are in the form of **paired data** (x, y) which are obtained from a **common source**. The x-values are from the first population and the y-values are from the second population, or else two sets of data values are selected from the same population.

3. A **Point estimator of** $\mu_1 - \mu_2$ -

 (a) For independent sampling, the point estimate of $\mu_1 - \mu_2$ is $\bar{x}_1 - \bar{x}_2$ where $\bar{x}_1$ and $\bar{x}_2$ are the sample means from sample 1 and sample 2, respectively.
 (b) For dependent sampling, the point estimate of $\mu_1 - \mu_2 = \mu_d$ is the sample mean $\bar{d}$ of the differences $x - y$ taken from the paired samples (x, y).

4. **Test statistic** - a z-value or a t-value generated from the point estimate of $\mu_1 - \mu_2$. When the test statistic is in the **critical region** for the test conducted, the **null hypothesis** is rejected.

5. **Confidence interval for** $\mu_1 - \mu_2$ **with confidence level** $(1 - \alpha)*100\%$ - an interval whose midpoint is the point estimate. Before the sample has been chosen, there is a probability of $1 - \alpha$ that $\mu_1 - \mu_2$ will fall inside this interval. The distance from the midpoint to an end of the interval is the **maximum error of the estimate** E. The most common confidence levels are 90%, 95%, and 99%.

6. **The prob-value (or p-value) for a hypothesis test -** A value p generated by the point estimate such that rejection of the null hypothesis occurs whenever $p \leq \alpha$, where α is the level of significance of the test.

7. **Independent sample methods relating to $\mu_1 - \mu_2$ -** For small sample sizes ($n_1 < 30$ or $n_2 < 30$) or for large sample sizes ($n_1, n_2 \geq 30$), one of the following two methods can be employed for hypothesis tests and confidence intervals for $\mu_1 - \mu_2$:

(a) **Equal population variances**. Then a **pooled estimate** s_p for the common standard deviation is used:

$$s_p = \sqrt{\frac{(n_1 - 1)s_1^2 + (n_2 - 1)s_2^2}{n_1 + n_2 - 2}}$$

The test statistic becomes

$$t = \frac{(\bar{x}_1 - \bar{x}_2) - (\mu_1 - \mu_2)}{s_p \sqrt{\dfrac{1}{n_1} + \dfrac{1}{n_2}}}$$

with $df = n_1 + n_2 - 2$.

(b) **Unequal population variances**. The test statistic used is

$$t = \frac{(\bar{x}_1 - \bar{x}_2) - (\mu_1 - \mu_2)}{\sqrt{\dfrac{s_1^2}{n_1} + \dfrac{s_2^2}{n_2}}} \quad \text{with} \quad df = \frac{\left(s_1^2/n_1 + s_2^2/n_2\right)^2}{\dfrac{\left(s_1^2/n_1\right)^2}{n_1 - 1} + \dfrac{\left(s_2^2/n_2\right)^2}{n_2 - 1}}.$$

PROCEDURES

In this lab session, **MINITAB** will be used to produce statements concerning hypothesis tests and confidence intervals for the difference of two population means. Load **MINITAB** according to the procedure discussed in *Lab #0*.

1. THE DIFFERENCE OF TWO   POPULATION MEANS USING DEPENDENT SAMPLES

When two populations are generated from a **common source** resulting in pairs of the form (x, y), the x-values come from Population 1 and the y-values from Population 2. In order to produce a hypothesis test result or generate a confidence interval, **dependent sampling** is performed on the n data pairs $(x_1, y_1), (x_2, y_2), ..., (x_n, y_n)$. By forming the **population of differences** $d = x - y$ from all ordered pairs (x, y) with this common source, hypothesis tests and confidence intervals for the difference $\mu_1 - \mu_2$ will be reduced to the 1-sample methods discussed previously in *Lab #8* and *Lab #9*. The notation μ_d will be used instead of $\mu_1 - \mu_2$. Each difference from a sample data pair (x_k, y_k) will have the form $d_k = x_k - y_k$. The **point estimate** of μ_d is the sample mean $\bar{d}$ of the differences $d_1, d_2, ..., d_n$.

194

Example 1: Suppose that the backers of the *StaFit* weight-loss program would like to advertise that their female customers lose, on average, more than 20 pounds over the six week course period. To test this claim, at a level of significance of 0.05, 20 females who participated in the program were randomly selected. Their **before program** weights, **after program** weights, and **weight loss** (in pounds) are recorded in the following table:

Person Number	1	2	3	4	5	6	7	8	9	10
Before Weight	156	181	205	162	159	192	177	215	160	173
After Weight	132	150	162	137	130	166	152	175	150	166
Weight Loss	24	31	43	25	29	26	25	40	10	7

Person Number	11	12	13	14	15	16	17	18	19	20
Before Weight	160	231	158	167	153	175	172	141	164	193
After Weight	133	181	142	149	131	143	155	130	146	164
Weight Loss	27	50	16	18	22	32	17	11	18	29

In the **MINITAB Data** window, place labels of *Before*, *After*, and *Wt Loss* on columns C1, C2, and C3, respectively. Type in the *Before Weight* values in column C1 and the *After Weight* values in column C2.

To produce the **weight loss** values in column C3, select **Calc→Mathematical Expressions**, type *C3* in the **Variable** box, and type *C1-C2* in the **Expression** box. After **OK** is clicked, one will find the weight loss values in column C3.

Population 1 will represent the **before program** weights and **Population 2** the **after program** weights of all participants who completed the program.

The hypothesis test of interest is the right-tail test given by:

$$H_0: \mu_d = 20 \ (\leq) \qquad \text{vs.} \qquad H_1: \mu_d > 20 \qquad \text{with level of significance } \alpha = 0.05$$

In order to justify using a 1-sample t-test of the weight losses in column C3, we must see if the population of weight losses is approximately normally distributed.

As indicated earlier, a test of population normality can be conducted by selecting **Graph→Normal Plot**. When the **Anderson-Darling Normality Test** is applied to column C3, the p-value is 0.649 for the hypothesis test:

H_0: the population is normal vs. H_1: the population is not normal

This demonstrates that we are justified in assuming that the population of weight losses of females in the program are approximately normally distributed.

A simplistic approach is to construct histogram of the sample of weight losses. Applying **Graph→Histogram** to column C3 results in the histogram in figure *Minitab 10.1*. Observe that the shape of this histogram is consistent with the assumption that the population of weight losses is approximately normally distributed.

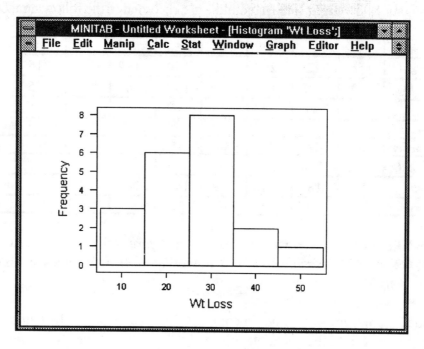

Minitab 10.1

In order to reach a conclusion for the weight loss hypothesis test, select **Calc→Basic Statistics→1-Sample t** and fill in the dialog box as indicated in figure *Minitab 10.2*.

Minitab 10.2

196

After **OK** has been clicked, the **Session** window shown in figure *Minitab 10.3* will reveal the steps taken for producing the **Anderson-Darling Normality Test** and the histogram. Information is also displayed relating to the 1-sample t-test on column C3.

```
┌─────────────────────────── Session ───────────────────────────┐
│ MTB > %NormPlot 'Wt Loss'.                                     │
│ Executing from file: C:\MTBWIN\MACROS\NormPlot.MAC             │
│ Macro is running ... please wait                              │
│ MTB > Histogram 'Wt Loss';                                    │
│ SUBC>   MidPoint;                                             │
│ SUBC>   Bar;                                                  │
│ SUBC>   Title "";                                            │
│ SUBC>   Axis 1;                                              │
│ SUBC>   Axis 2.                                              │
│ MTB > TTest 20.0 c3;                                          │
│ SUBC>   Alternative 1.                                        │
│                                                               │
│ TEST OF MU = 20.00 VS MU G.T. 20.00                          │
│                                                               │
│              N     MEAN    STDEV   SE MEAN       T   P VALUE  │
│ Wt Loss     20    25.00    10.99      2.46    2.04    0.028   │
│                                                               │
│ MTB >                                                         │
└───────────────────────────────────────────────────────────────┘
```

Minitab 10.3

Since the p-value from the **Session** window is shown to be $P = 0.028 \leq 0.05$, where $\alpha = 0.05$ is the stated level of significance of the test, there is sufficient evidence to conclude that the mean weight loss from the program is greater than 20 pounds. The backers of the weight-loss program are justified in advertising their average weight loss claim.

Observe that the **Session** window information displayed in figure *Minitab 10.3* also reveals the test statistic $t = 2.04$. As discussed in *Lab #8*, **MINITAB** could have been used to determine the critical value for the right tail of the rejection region by selecting **Calc→ Probability Distributions→ T**, followed by clicking on **Inverse cumulative probability**.

This critical value for a degrees of freedom $df = 19$ is 1.7291, and since $2.04 > 1.7291$, the same conclusion is obtained as above, namely to reject the null hypothesis.

In order to have **MINITAB** produce a **99% confidence interval** of the difference μ_d in the means of the population of "before" and "after" weights, select **Stat→Basic Statistics→1-Sample t**, click on **Confidence interval**, type *C3* in the **Variables** text box, and type *99.0* in the **Level** text box.

After **OK** is clicked, the **Session** window will display the desired confidence interval information, as shown in figure *Minitab 10.4*.

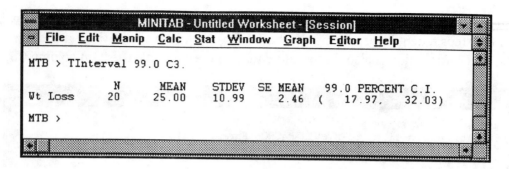

Minitab 10.4

The confidence interval (17.97, 32.03) can be interpreted as follows:

*The backers of the **StaFit** program are 99% confident that the average weight loss of its participants is between 17.97 pounds and 32.03 pounds.*

2. THE DIFFERENCE OF TWO POPULATION MEANS: LARGE OR SMALL INDEPENDENT SAMPLES

In comparing two population means a common procedure for data collection is to produce two **independent samples** as mentioned in the **BACKGROUND INFORMATION**. Most basic statistics texts separate the testing methods into **large sample sizes** (both at least 30) and using the z-statistic, as opposed to **small sample sizes** (one or more less than 30) and using the t-statistic. It is not necessary to make such a distinction, since the results will be more reliable when using the t-statistic in all cases.

The **MINITAB** menu options **Stat→Basic Statistics→2-Sample t** are used to either form a conclusion relating to a hypothesis test of $\mu_1 - \mu_2$ or generate a confidence interval for $\mu_1 - \mu_2$. It is assumed that independent sampling is conducted from the two populations. The two samples from the populations are to be entered in columns in the **Data** window; the **MINITAB** conclusions will appear in the **Session** window.

Observe that there is no **2-Sample Z** test in **MINITAB** as in the 1-population situation when the population variance is known. As mentioned above, both large and small samples will be treated in the same manner.

In many instances, the two population means μ_1 and μ_2 are to be compared by using independent sampling where, for practical reasons such as cost and effort, small samples from one or both populations are necessary. In this event it is necessary that the two populations are approximately normally distributed.

As mentioned in **BACKGROUND INFORMATION** part of this session, two methods can be employed.

(a) Method 1 - Equal Population Variances

A preliminary test of this statement could be set up, such as:

$H_0:$ $\sigma_1^2 = \sigma_1^2$ vs. $H_1:$ $\sigma_1^2 \neq \sigma_1^2$ **level of significance = $\alpha = 0.05$**

This test involves the **F-distribution**, which will not be discussed here. If one fails to reject the null hypothesis, then **Method 1** can be employed.

Warning: The above F-test is highly sensitive to any lack of normality of the two populations in which case the results may not be reliable. Instead, one may use the following **rule of thumb:**

i. there are no outliers
ii. the two standard deviation differ from each other by no more than a factor of two.

This method involves using a **pooled variance** s_p for the two samples, which enables one to use a degrees of freedom $df = n_1 + n_2 - 2$ on a Student's t distribution. The formulas for the polled variance and Student's t-statistic are:

$$t = \frac{(\bar{x}_1 - \bar{x}_2) - (\mu_1 - \mu_2)}{s_p\sqrt{\dfrac{1}{n_1} + \dfrac{1}{n_2}}} \quad \text{and} \quad s_p = \sqrt{\frac{(n_1 - 1)s_1^2 + (n_2 - 1)s_2^2}{n_1 + n_2 - 2}}$$

The advantage of this method is that a smaller confidence interval or a smaller p-value will usually result than when using the more general procedure discussed below in **Method 2**.

(b) Method 2 - Unequal Population Variances (general method)

If one concludes that the two population variances are unequal, then **Method 2** should be employed. The formula for the Student's t-statistic and degrees of freedom are given by the following formulas:

$$t = \frac{(\bar{x}_1 - \bar{x}_2) - (\mu_1 - \mu_2)}{\sqrt{\dfrac{s_1^2}{n_1} + \dfrac{s_2^2}{n_2}}} \quad \text{and} \quad df = \frac{\left(s_1^2/n_1 + s_2^2/n_2\right)^2}{\dfrac{(s_1^2/n_1)^2}{n_1 - 1} + \dfrac{(s_2^2/n_2)^2}{n_2 - 1}}$$

Method 2 is a more conservative approach relating to hypothesis tests and confidence intervals. The confidence intervals will generally have a larger width and it will generally be more difficult to reject a null hypothesis relating to $\mu_1 - \mu_2$.

Example 2: Load the **MINITAB worksheet** data file called *lake.mtw* into the **Data** window. Refer to *Lab #0* for instructions on how to carry this out.

The **Data** window contains five columns of information relating to 71 lakes in a certain geographic region. In columns C2 and C3 you will find the average lake depths (in feet) and average pH readings, respectively. Suppose that we are interested in comparing the average pH reading of "deep" lakes with the average pH reading of "shallow" lakes. For our purposes we will define a "shallow" lake to have an average depth of less that 30 feet and a "deep" lake to have an average depth of at least 30 feet. If pH < 7 the lake is acidic; if pH > 7 it is basic; if pH $= 7$ (such as pure water), it is neutral.

In order to efficiently sort the pH data located in column C3 by average lake depth, label column C6 as *Sort D* and column C7 as *Sort pH*. After selecting **Minip→Sort**, fill in the **Sort** dialog boxes as indicated in figure *Minitab 10.5*.

Minitab 10.5

Column C6 will contain the 71 average lake depths, listed from smallest to largest while column C7 will contain the average pH values of the lakes corresponding to the column C6 values.

In order to separate the pH values into two columns corresponding to shallow and deep lake, label columns C8 and C9 as *Shallow* and *Deep*, respectively. Use the mouse to highlight the first 38 pH values found in column C7 (shallow lakes). Apply **Edit→ Copy Cells** and then after moving to Column C8, apply **Edit →Paste/Replace Cells**.

In the same manner, paste the last 33 pH values in column C7 (deep lakes) into column C9. The 38 pH levels for the shallow lakes and 33 pH levels for the deep lakes are now located in columns C7 and C8, respectively.

One may assume that the two sets of pH levels have been obtained by independent sampling. Before conducting a hypothesis test or producing a confidence interval from two samples, it is of benefit to gather some statistical information such as obtaining the sample mean and sample standard deviation as well as observing if their shapes are approximately normal.

By applying **Stat→Basic Statistics→Descriptive Statistics**, figure *Minitab 10.6* provides us with the following information about these columns from the **Session** window.

```
                                    Session
MTB > Describe   'Shallow' 'Deep'.

                N     MEAN    MEDIAN   TRMEAN    STDEV   SEMEAN
Shallow        38    6.813    6.900    6.765    0.714    0.116
Deep           33    7.103    7.000    7.072    0.614    0.107

               MIN     MAX      Q1       Q3
Shallow       5.700   8.800   6.200    7.225
Deep          6.000   8.600   6.800    7.400

MTB >
```

Minitab 10.6

Suppose that it is claimed that the shallow lakes have a higher average pH level (and thus are more acidic) than the deep lakes. This claim will be tested with a hypothesis test of the difference of the two mean using a level of significance of 0.05. Here is the setup for this left-tail test:

$$H_0: \mu_1 - \mu_2 = 0 \ (\geq) \quad \text{vs.} \quad H_1: \mu_1 - \mu_2 < 0 \quad \text{with } \alpha = 0.05$$

Population 1 and **Population 2** for this test correspond to the shallow and deep lakes pH levels, respectively.

In will now be necessary to investigate whether or not the two populations of pH values are approximately normally distributed. By choosing **Calc→Histogram, connect** in the display box, and **Overlay graphs on the same page** in the **Frame** box, figure *Minitab 10.7* provides us with information about the shapes of the shallow lake pH levels and the deep lake pH levels.

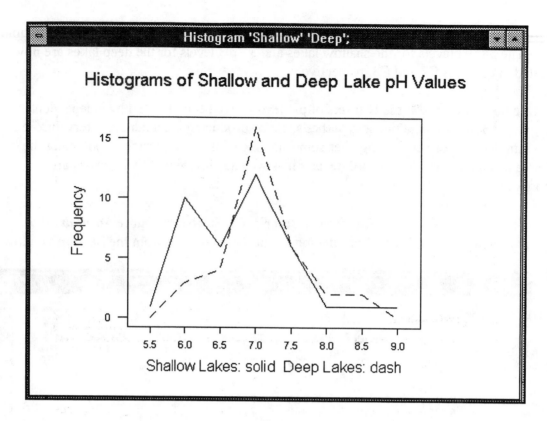

Minitab 10.7

The two samples of pH levels appear to be approximately normally distributed. By applying the **Anderson-Darling Normality Test** on columns C8 and C9 after choosing **Graph→Normal Plot**, one obtains p-values of 0.207 and 0.122, respectively. Since

H_0: **"the population is normal" vs. H_1: "the population is not normal"**

is the hypothesis test of normality, the above p-values provide conclusive evidence that the shallow lake and deep lake pH levels are approximately normally distributed.

MINITAB will now enable us to state a conclusion as to whether or not the shallow lakes are, on average, more acidic than deep lakes. Select **Stat→Basic Statistics→ 2-Sample t** and fill in the **2-Sample t** dialog box as shown in figure *Minitab 10.8*.

By comparing the two sample standard deviations of columns C8 and C9 in figure *Minitab 10.6* and observing the shapes of the two histograms in figure *Minitab 10.7*, it is reasonable to assume equal population variances. For this reason, the **Assume equal variances** box was checked. **Less than** was selected in the **Alternative** box since we are conducting a left-tail test.

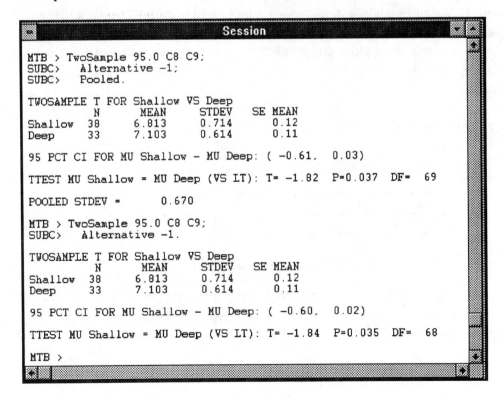

Minitab 10.8

This procedure can also be carried out assuming unequal population variances. After carrying out both procedures of an equal variance and unequal variance test, the **Session** window will provide us with the information as displayed in figure *Minitab 10.9*.

```
MTB > TwoSample 95.0 C8 C9;
SUBC>   Alternative -1;
SUBC>   Pooled.

TWOSAMPLE T FOR Shallow VS Deep
            N       MEAN      STDEV    SE MEAN
Shallow    38      6.813      0.714     0.12
Deep       33      7.103      0.614     0.11

95 PCT CI FOR MU Shallow - MU Deep: ( -0.61,  0.03)

TTEST MU Shallow = MU Deep (VS LT): T= -1.82  P=0.037  DF=  69

POOLED STDEV =      0.670

MTB > TwoSample 95.0 C8 C9;
SUBC>   Alternative -1.

TWOSAMPLE T FOR Shallow VS Deep
            N       MEAN      STDEV    SE MEAN
Shallow    38      6.813      0.714     0.12
Deep       33      7.103      0.614     0.11

95 PCT CI FOR MU Shallow - MU Deep: ( -0.60,  0.02)

TTEST MU Shallow = MU Deep (VS LT): T= -1.84  P=0.035  DF=  68

MTB >
```

Minitab 10.9

The p-value by assuming equal population variances is $p = 0.037$ whereas for unequal population variances we have $p = 0.035$. The assumption of unequal population variances is a more conservative procedure than for equal population variances. Consequently, one usually obtains a smaller p-value assuming equal population variances, contrary to the results of the current test.

Since each p-value is less than the level of significance $\alpha = 0.05$, we are justified in saying:

"At a level of significance of 0.05, there is sufficient evidence to conclude that the mean pH level of the deep lakes is greater than the mean pH level for the shallow lakes."

We can then conclude, at a level of significance of 0.05, that on average the shallow lakes are more acidic than the deep lakes.

Observe from figures *Minitab 10.8* and *Minitab 10.9* that we are also producing a 95% confidence interval for the difference of the population means $\mu_1 - \mu_2$. The 95% confidence interval obtained assuming equal population variances was found to be $(-0.61, 0.03)$. This can be interpreted by the following sentence:

"At a 95% level of confidence, we can state that $-0.61 < \mu_1 - \mu_2 < 0.03$."

Example 3: An automobile testing agency is interested in determining if a major automotive company is currently producing more fuel efficient vehicles as compared to those five years ago. A random sample of 20 new vehicles were tested for fuel efficiency five years ago and another random sample of 20 new vehicles were tested this year.

The information is summarized in the following table. Data is given in miles per gallon (MPG) under normal driving conditions.

Fuel Efficiency - Five Years Ago

17	22	28	23	21	25	27	31	18	20
19	24	26	19	23	18	25	21	27	18

Fuel Efficiency - Current Year

29	32	27	24	21	26	23	18	24	27
21	20	27	23	33	29	27	25	22	30

Let **Population 1** represent the MPG s of the vehicles five years ago and **Population 2** the MPG s of the vehicles currently produced. Label columns C1 and C2 as *Old MPG* and *New MPG*, respectively. Then enter the 20 five year ago data values in column C1 and the 20 current data values in column C2.

It will now be necessary to apply a **2-sample t-test** of the difference $\mu_1 - \mu_2$ of the population means. First, let us test the two populations for normality. Figure *Minitab 10.10* reveals that the shapes of the distributions of the two samples appear normally distributed. This figure results from the selection of **Graph→Histogram** and **Frame→Multiple Graphs→Overlay graphs on the same page**.

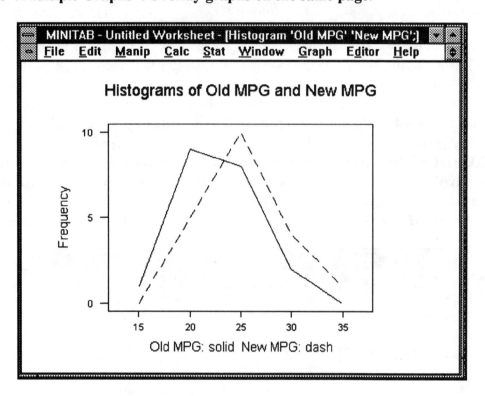

Minitab 10.10

To test the hypotheses **H_0: "the Old MPG population is normal"**, select **Graph→Normal Plot** and apply the **Anderson-Darling Normality Test**. The p-value for the *Old MPG* column is 0.512 which means that we can assume the population of *Old MPG* values is approximately normally distributed. Similarly, when this test is applied to the *New MPG* values, the resulting p-value is 0.909. This is a strong indicator that the population of *New MPG* values are approximately normally distributed.

Next, we will determine if it reasonable to assume that the two population variances are equal. For this purpose, use **MINITAB** to compare the two sample standard deviations of columns C1 and C2. After selecting **Stat→Basic Statistics→Descriptive Statistics**, figure *Minitab 10.11* displays the **Session** window results.

```
┌─────────────────────────────────────────────────────────────────────────┐
│ ─                           Session                              ▼  ▲     │
│ MTB > Describe 'Old MPG' 'New MPG'.                                  ▲    │
│                                                                          │
│                   N      MEAN    MEDIAN    TRMEAN     STDEV    SEMEAN     │
│  Old MPG         20     22.600   22.500    22.444     3.952    0.884      │
│  New MPG         20     25.400   25.500    25.389     4.031    0.901      │
│                                                                          │
│                  MIN      MAX        Q1        Q3                         │
│  Old MPG      17.000   31.000    19.000    25.750                        │
│  New MPG      18.000   33.000    22.250    28.500                    ▼    │
│ ◄ ►                                                               ► │
└─────────────────────────────────────────────────────────────────────────┘
```

Minitab 10.11

Observe that the two standard deviations of 3.952 miles per gallon for the *Old MPG* data and 4.031 for the *New MPG* data are nearly equal to each other.

Outliers in each of the two data sets can easily be checked with **MINITAB** by producing a **boxplot**. Any outlier is represented by an asterisk (*). Recall that an outlier is any data value less than $Q_1 - 1.5 *IQR$ or greater than $Q_3 + 1.5 *IQR$ (see **BACKGROUND INFORMATION** in *Lab #1*) where IQR is the **interquartile range** given by $IQR = Q_3 - Q_1$. After selecting **Graph→Boxplot** and **Frame→Multiple Graphs→Overlay graphs on the same page**, a boxplot of columns C1 and C2 results as shown in figure *Minitab 10.12*. We must also select the option to display the outliers.

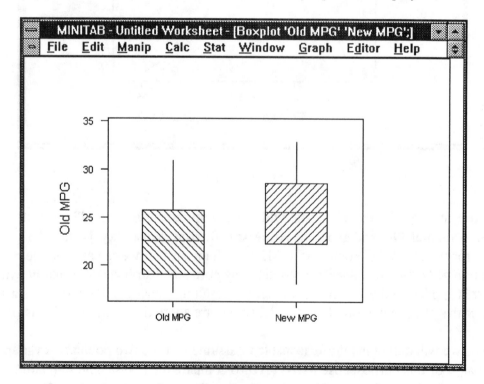

Minitab 10.12

Notice that the absence of an asterisk indicates that there are no outliers in columns C1 and C2.

It should be clear by now that we are justified in assuming that the two populations are approximately normally distributed and that the population variances are equal. Consequently, **Method 1** can be utilized to test the hypothesis of an increased average miles per gallon for the current vehicles from the average five years ago.

We now set up the following hypothesis test (left tail):

$$H_0: \mu_1 - \mu_2 = 0 \ (\geq) \qquad \text{vs.} \qquad H_1: \mu_1 - \mu_2 < 0 \qquad \textbf{level of significance: } \alpha = 0.05$$

Select **Stat→Basic Statistics→2-Sample t** and fill the text boxes as shown in figure *Minitab 10.13*.

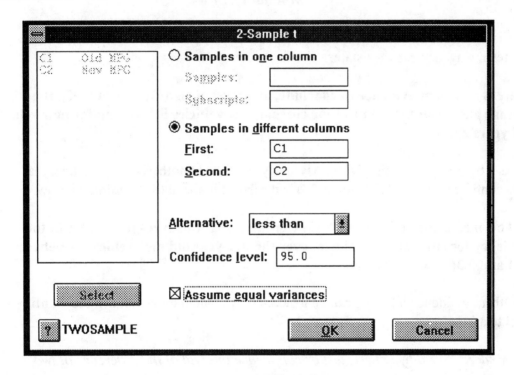

Minitab 10.13

Observe that **Samples in different columns** is chosen and the **Assume equal variances** box is checked. Since a left-tail test is to be conducted, **less than** is chosen in the **Alternative** box.

After the **OK** button is clicked, the **Session** window will reveal the information as displayed in figure *Minitab 10.14*.

```
┌─────────────────────────────── Session ─────────────────────────────┐
│ MTB > TwoSample 95.0 C1 C2;                                          │
│ SUBC>    Alternative -1;                                             │
│ SUBC>    Pooled.                                                     │
│                                                                      │
│ TWOSAMPLE T FOR Old MPG VS New MPG                                   │
│             N      MEAN      STDEV     SE MEAN                        │
│ Old MPG   20      22.60      3.95       0.88                         │
│ New MPG   20      25.40      4.03       0.90                         │
│                                                                      │
│ 95 PCT CI FOR MU Old MPG - MU New MPG: ( -5.36,   -0.24)             │
│                                                                      │
│ TTEST MU Old MPG = MU New MPG (VS LT): T= -2.22   P=0.016   DF=  38  │
│                                                                      │
│ POOLED STDEV =        3.99                                           │
│                                                                      │
│ MTB > |                                                              │
└──────────────────────────────────────────────────────────────────────┘
```

Minitab 10.14

The p-value is P = 0.016 which is less than $\alpha = 0.05$ so the null hypothesis is rejected. The testing agency can thus state:

There is sufficient evidence to conclude, at a level of significance of 0.05, that the average gas consumption rates for current new vehicles is less than for new vehicles five years ago.

As a by-product of this **MINITAB** analysis of the hypothesis test, observe that a 95% confidence interval (-5.36, -0.24) is displayed in the above **Session** window.

This means that one can be 95% confident that the average increase in fuel efficiency for current new vehicles over the five year old new vehicles is between 0.24 and 5.36 miles per gallon.

Other confidence intervals can be generated by changing the value in the **Confidence level** text box.

Reminder: *For a hypothesis test or confidence interval of the difference of two population means where the population variances are assumed to be different, do not check the* **Assume equal variances box**.

LAB #10: *DATA SHEET*

Name: _____   *Date:* _____

Course #: _____   *Instructor:* _____

1. Suppose that the advertising department for a shoe manufacturer believes that the soles of their premium running shoes (*Brand A*) last longer than a competitor's premium running shoes (*Brand B*).   Runners are chosen to test a shoe of each brand for sole wear. Tread wear on a shoe is measured in miles. Here were the results of the test on 20 paired samples:

Runner No.	1	2	3	4	5	6	7	8	9	10
Brand A	855	930	945	875	660	945	1010	745	880	970
Brand B	760	890	910	900	635	885	995	830	865	980

Runner No.	11	12	13	14	15	16	17	18	19	20
Brand A	745	790	995	1020	885	595	770	885	920	990
Brand B	665	830	865	935	870	610	725	840	875	970

(a) Explain why the dependent sample method has been used; identify the common source. Also, explain what controls must be placed in order to reduce sampling bias.

(b) Store the *Brand A* data values in column C1 (label: *Brand A*) and the *Brand B* data values in column C2 (label: *Brand B*). Then use **MINITAB** to produce the *Brand A - Brand B* data values in column C3 (label: *Diff*). Provide a printout of the **Data** window containing the three columns of values. *Be careful to highlight only the cells relating to these columns!*

209

(c) Set up a hypothesis test concerning the advertising department's claim. Use a level of significance of 0.05.

Hypothesis Test: H$_0$: _____. H$_1$: _____.

(d) Apply the **Stat→Basic Statistics→1-Sample t** commands in order to produce the information relating to this hypothesis test. Provide a printout of the **Session** window of the results obtained. State a well-formed conclusion relating to the advertising department's claim.

Conclusion:

(e) Generate a 95% confidence interval for the difference of the means of the *Brand A* and *Brand B* sole lives. Provide a printout of the appropriate part of the **Session** window. Interpret your results.

(f) Produce a histogram of the sole life differences in column C3. Also, apply the Anderson-Darling Normality Test to C3 to determine whether or not to reject the null hypothesis **H$_0$: "the population is normal"**, at a level of significance of 0.05. Based on this information, do you believe that you were justified in carrying out the 1-Sample t-test? Explain.

2. A school system wishes to have a hypothesis test run to see if there is any change in the standardized test scores of their high school seniors for 1995 versus 1990, in light of the innovative changes made in teaching and administrative structure. Random scores of 18 seniors from 1990 and 24 students from 1995 are displayed in the following tables:

Senior Scores for 1990

69	66	43	81	73	84	91	66	72
75	49	82	66	69	73	82	80	51

Senior Scores for 1995

72	85	92	83	74	85	66	62	85	83	77	68
94	61	72	96	86	55	71	71	89	83	76	90

(a) Explain why independent sampling is called for in this situation, as opposed to dependent sampling?

(b) Set up a hypothesis test on the basis of "a change" in the average test scores of the two classes of seniors. Run the test at a 0.05 level of significance. Use **MINITAB** to produce the information needed for this hypothesis test. Printout the **Session** window containing this information and state a conclusion by wording in the context of this problem.

Hypothesis Test: H_0: _____. H_1: _____.

Conclusion:

(c) Store the 18 senior grades from 1990 in column 1 (label: *Senior 1*) and the 24 senior grades from 1995 in column 2 (label: *Senior 2*). Assuming unequal population variances, use **MINITAB** to generate information in the **Session** window relating to this test. Provide a printout of the relevant part of this **Session** window. State an appropriate conclusion by wording in the context of this problem.

Conclusion:

(d) Repeat part (c) by assuming equal population variances. What differences have you noticed in the **Session** window this time as opposed to the **Session** window in part (c)?

(e) Produce a histogram containing both the column C1 and column C2 data. Label the axes appropriately. Also, apply the **Anderson-Darling Normality Test** to both C1 and C2 to determine whether or not to reject the null hypothesis H_0: **"the population is normal"**, at a level of significance of 0.05. Based on this information, do you believe that you were justified in carrying out the **2-Sample t-test**? Explain.

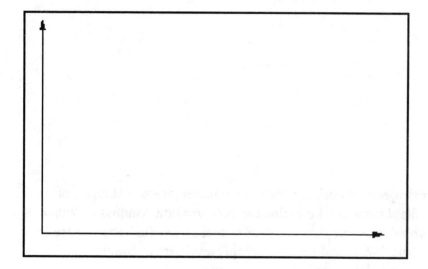

213

(f) Test for equal population variances by comparing the standard deviation of C1 with the standard deviation of C2 after applying **Stat→Basic Statistics→Descriptive Statistics**. Print out the **Session** window containing the information produced from this command. Also, check for outliers in C1 and C2 by generating and printing out a boxplot. See figures *Minitab 10.10*, *Minitab 10.11*, and *Minitab 10.12* for the procedures required to carry out the above steps. Are you justified in assuming that the population variances are equal? Explain.

3. **Team Exploration Project.** Consider *Example 2* relating to the pH levels of the 71 lakes found in the data file *lake.mtb*. Load this file into the **Data** window by following the procedure from *Lab #0*. Observe that the areas of these lakes (in hectares) are to be found in Column C1 (label: *AREA*) and the average pH levels are in column C3 (label: *PH*). Turn in a detailed report of your observations.

(a) Consider a lake to be **small** if the number of hectares is below 150, which is approximately 0.579 square miles. All other lakes will be called **large**. Use the procedure as discussed in figure **Minitab 10.5** to sort the pH levels according to size, namely **small** vs. **large**. Use column C6 (label: *Sort A*) for the sorted areas, column C7 (label: *Sort pH*) for the pH values of the sorted areas, column C8 (label: *Small*) for the pH levels of small lakes, and column C9 (label: *Large*) for the pH levels of the large lakes. Turn in a printout of the **Data** window of columns C1 through C9. *Be careful to highlight only the data cells for these columns!*

(b) Let **Population 1** and **Population 2** represent the pH levels of the **small** lake pH levels and **large** lake pH levels, respectively. Use **MINITAB** to construct a histogram of the two data samples in columns C8 and C9, in the manner of figure *Minitab 10.7*. Discuss the shapes of these two histograms related to population normality. Provide a printout of the histogram.

(c) Produce **Anderson-Darling Normality Test** graphs for columns C8 and C9 by applying **Graph→Normal Plot**. Based on the resulting p-values, determine whether or not one can assume that the two populations are approximately normally distributed. The null hypothesis should be **H₀ : "the population is normal"**. Let the level of significance be α = 0.05. Provide a printout of the two normal plot displays.

(d) Compare the two sample standard deviations of C8 and C9 by applying **Stat→ Basic Statistics→Descriptive Statistics**. Provide a printout of the **Session** window contents of the descriptive statistics information obtained.

(e) Construct a hypothesis test, at a level of significance of 0.05, to determine whether the mean pH level of the small lakes is less than the mean pH level of the large lakes. In **MINITAB** select **Stat→Basic Statistics→2-Sample t** and use both **Method 1** (equal population variances) and **Method 2** (unequal population variances). Provide a display of the **Session** window containing the results of these tests. State a well-formed conclusion related to the pH levels of the lakes by using the p-value information in the **Session** window.

(f) Use MINITAB to generate a 99% confidence interval of the difference $\mu_2 - \mu_1 =$ mean pH level for **large** lakes - mean pH level for **small** lakes. Provide a printout of the **Session** window results. Interpret your results.

4. **Team Exploration Project**. There are numerous examples of hypothesis tests and confidence intervals that involve dependent sampling procedures. The common source is often a person - "before" data versus "after" data. Other examples are plant growth from time 1 to time 2 and comparing two brands by paring them by common controls.

Design your own experiment involving two population means using **dependent** sampling. Decide on the type of hypothesis test to use to compare the two population means, and the level of significance to use. Determine, based on the differences of the data pair values whether or not it is appropriate to use the 1-sample t test. Apply **MINITAB** in order to form a well-stated conclusion. Provide appropriate printouts of the session window. Write up a detailed report of the project.

5. **Team Exploration Project**. Design your own experiment involving two population means using independent sampling. Determine whether the assumption of approximately normal populations is satisfied by using various methods. Decide on the type of hypothesis test to use to compare the two population means, sample sizes from the two populations, and state an appropriate level of significance. Decide on whether the assumption of equal population variances is appropriate in this situation. Apply **MINITAB** in order to form a well-stated conclusion. Provide appropriate printouts of the session window. Turn in a detailed report of the project.

NOTES

STATISTICS LAB # 11

CHI-SQUARE GOODNESS-OF-FIT TESTS

PURPOSE - to use MINITAB to

1. apply the *chi-square goodness-of-fit* test to *multinomial experiments*
2. apply the *chi-square goodness-of-fit* test to *contingency tables*

BACKGROUND INFORMATION

1. In the *multinomial experiment*, suppose that we have k populations. Let P_i be the probability that a value is in population i ($i = 1, 2, ..., k$). Then, for specified values p_i, the null hypothesis of interest will be H_0: $P_i = p_i$ for **all** i against the alternative H_1: $P_i \neq p_i$ for **some** i.

2. To test the null hypothesis in (1), select a random sample of size n and let $n = N_1 + N_2 + ... + N_k$ where N_i is the sample size from population i.

3. If the null hypothesis is true, then a value from population i will have a probability p_i of occurring. Thus, the expected value is E[*population i*] $= n\,p_i$ which we denote by e_i.

4. The *test statistic* for this test is $\chi^2 = \sum \dfrac{(N_i - e_i)^2}{e_i}$ with df = k - 1.

5. The null hypothesis is rejected if the test statistic is **greater than or equal to** the critical value (table value) of $\chi^2_{k-1,\alpha}$, for a given significance level α.

6. When we consider *contingency tables*, we will consider a *test for independence in populations when the data values are classified according to two characteristics*.

7. Let the two characteristics be denoted by X and Y. Let the possible values for the X characteristic be denoted as 1, 2, 3, ..., r and the possible values for the Y characteristic be denoted by 1, 2, 3, ..., s.

8. Let P_{ij} denote the proportion of the population that has both X characteristic i and Y characteristic j. Let P_i be the proportion of the population with X characteristic i and Q_j be the proportion with Y characteristic j.

9. The *null hypothesis* for the *test of independence* under these conditions for the contingency table is H_0: $P_{ij} = P_i Q_j$ for all $i = 1, 2, ..., r$ and $j = 1, 2, ..., s$. The *alternative hypothesis* is H_1: $P_{ij} \neq P_i Q_j$ for some values of i and j.

10. Let N_{ij} be the number of elements of the sample with both X characteristic i and Y characteristic j and let n be the sample size. The expected value of N_{ij} is then given by $e_{ij} = n P_i Q_j$. In practice, we do not know the values of P_i and Q_j, so we will have to estimate e_{ij}. We compute e_{ij} from $e_{ij} = \dfrac{(Row\ i\ total)(Column\ j\ total)}{Sample\ size}$.

11. The *test statistic* for the test of independence for the contingency table is given by

$$\chi^2 = \sum\sum \frac{(N_{ij} - e_{ij})^2}{e_{ij}}$$ where the summations are over all i and j with

    df = (r - 1)(s - 1).

12. The null hypothesis is rejected if the test statistic is *greater than or equal* to the critical value (table value) of $\chi^2_{(r-1)(s-1),\alpha}$, for a given significance level α.

13. *Note:* In both of the above two tests, you can apply the *p-value* approach to test the hypotheses.

PROCEDURES

First, load the **MINITAB** (windows version) software as in *Lab #0*.

1. THE CHI-SQUARE GOODNESS-OF-FIT TEST APPLIED TO A MULTINOMIAL EXPERIMENT

Example 1: A random sample of 200 college students who are smokers at a certain campus yielded the following breakdown in age groups. Also listed are the associated probabilities of a smoker being in an age group for that campus. These probabilities were obtained from past records on the smoking habits of students on this particular campus.

Age Group (years)	10 - 19	20 - 29	30 - 39	40 - 49
Frequency	10	107	60	23
Probability	0.04	0.46	0.40	0.1

Test whether there is a significant difference in the proportion of smokers for the different age groups. Use a significance level of $\alpha = 0.05$.

Note: **MINITAB** does not currently have a listed macro to do the chi-square goodness-of-fit test for the multinomial experiment. However, with a few simple commands you can do most of the computations for the test in **MINITAB**.

Let **Population 1** represent the age group 10 - 19, **Population 2** represent age group 20 - 29, etc. Then the appropriate null and alternative hypotheses are:

H₀: $P_1 = 0.04$, $P_2 = 0.46$, $P_3 = 0.4$, $P_4 = 0.1$

or

H₀: The proportion of smokers for the different age groups are as specified in the table.

H₁: At least one of these probabilities is different

or

H₁: The proportion of smokers for the different age groups are not as specified in the table.

In order to compute the expected frequencies from the information given, enter the observed frequencies of 10, 107, 60, and 23 in column C1. Enter the computed expected frequencies in column C2. The expected frequencies will be 8, 92, 80, and 20 for the respective age groups obtained by multiplying the probabilities by 200. Select **Calc→Mathematical Expressions** and the **Mathematical Expressions** dialog box will be displayed. In the **Variable** text box type *C3* and type the expression as shown in the **Expression** text box in figure *Minitab 11.1*. *Observe that this expression represents the test statistic.*

Minitab 11.1

Click on the **OK** button and a value of 8.39565 will be displayed in column C3 in the **Data** window. *If you wanted to save the test statistic as a constant such as K1 you could have typed K1 in the* **Variable** *text box and the computed value would have been saved as the constant K1. However, in order to see the value of K1 you would need to type PRINT K1 at the MTB> prompt in the* **Session window** *or select* **File→Display Data**.

To compute the *p-value* for the test, select **Calc→Probability Distributions→ Chisquare** and the **Chisquare Distribution** dialog box will be displayed. Select the **Cumulative probability** check box and type *3* (since $k = 4$ and df $= k - 1$) in the **Degrees of freedom** text box. Select the **Input column** check box and type *C3* in the corresponding text box. You may save in a column if you wish by specifying where to save in the **Optional storage** text box. The **Chisquare Distribution** dialog box with the appropriate selections and entries is shown in figure *Minitab 11.2*.

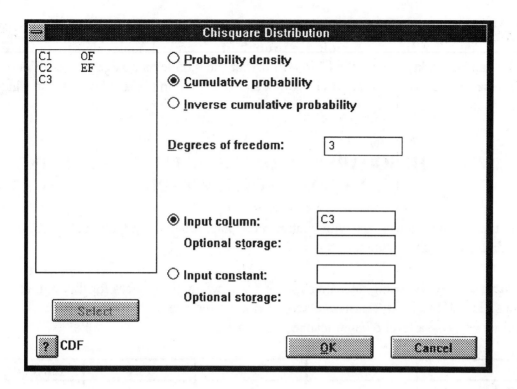

Minitab 11.2

Click on the **OK** button and the following **Session** window will be displayed as shown in figure *Minitab 11.3*.

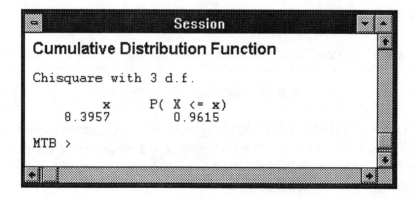

Minitab 11.3

From the above figure, $P(X \le x) = P(X \le 8.3957) = 0.9615$ will be displayed in the **Session** window since no storage cell was specified. Thus the
p-value = $1 - 0.9615 = 0.0385 < 0.05$, and so we will reject the null hypothesis that the proportion of smokers for the different age groups are as specified in the table.

Note: If you select **Inverse cumulative probability**, a critical (table) chi-square value of 7.8147 can be displayed in the **Session window** at the 0.05 level of significance. You will need to select the **Input constant** check box and type *0.95* in the text box. Again, since the test statistic value of 8.3957 > 7.8147, you will reject the null hypothesis.

2. THE CHI-SQUARE GOODNESS-OF-FIT TEST APPLIED TO A CONTINGENCY TABLE

In this section we will deal with contingency tables when the populations are classified according to two characteristics.

Example 2: The following table gives some TV viewing preferences for different ethnic groups. Use **MINITAB** to test whether viewing preferences and ethnicity are independent. Use a level of significance of $\alpha = 0.1$.

	CNN	*Cable*	*Pay*	*Early News*
White	272	566	299	147
Black	147	366	247	160
Spanish-speaking	192	421	266	131
Other	114	297	247	110

The appropriate null and alternative hypotheses are:

H₀: TV viewing preferences are independent of the ethnicity of the viewing group.

H₁: TV viewing preferences are not independent of the ethnicity of the viewing group.

Enter the table values in columns C1 through C4. Do not enter the row totals or column totals. Next select **Stat→Tables→Chisquare Test** and the **Chisquare Test** dialog box will be displayed. Select C1-C4 for the **Columns containing the table** text box. This is shown in figure *Minitab 11.4*.

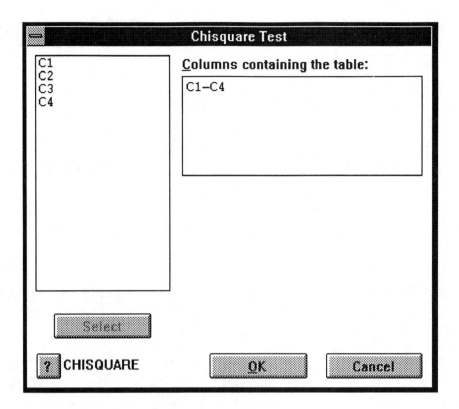

Minitab 11.4

Click on the **OK** button and the computations will be displayed in the **Session** window. This output is shown in figure *Minitab 11.5*. Observe that the computed test statistic value is 46.808 with a *p-value* of 0.000. Thus you would reject the null hypothesis at the 0.10 level of significance and claim that the TV viewing preferences are not independent of the ethnicity of the viewing group.

Note: (a) No column names were given since we needed a two-way classification. **MINITAB** automatically classified the rows as 1, 2, 3, and 4;

(b) A *p-value* of 0.000 implies very strong evidence that the variables are not independent.

```
=                           Session                           ▼ ▲

Chi-Square Test


Expected counts are printed below observed counts
            C1        C2        C3        C4      Total
    1      272       566       299       147      1284
         233.78    532.04    341.48    176.70

    2      147       366       247       160       920
         167.50    381.22    244.67    126.61

    3      192       421       266       131      1010
         183.89    418.51    268.61    139.00

    4      114       297       247       110       768
         139.83    318.23    204.25    105.69

Total      725      1650      1059       548      3982

ChiSq =   6.250 +   2.167 +   5.283 +   4.993 +
          2.510 +   0.607 +   0.022 +   8.806 +
          0.358 +   0.015 +   0.025 +   0.460 +
          4.771 +   1.417 +   8.949 +   0.176 = 46.808
df = 9,  p = 0.000

MTB >
```

Minitab 11.5

Note: The test for independence is unaffected by whether the marginal totals of one of the characteristics are fixed in advance or result from a random sample of the entire population. In this case, the test is called a test of homogeneity and the above null and alternative hypotheses will not be appropriate. The appropriate hypotheses will be H_0: The sampled populations are homogeneous vs. H_1: The sampled populations are not homogeneous.

LAB #11: *DATA SHEET*

Name: _____   *Date:* _____

Course #: _____   *Instructor:* _____

1. The following table gives the number of medals won in the 1992 Summer Olympics for the top six medal winning countries. The objective is to determine whether the type of medal that was won is independent of the country that won the medal.

	Gold	*Silver*	*Bronze*
Unified Team	45	38	29
United States	37	34	37
Germany	33	21	28
China	16	22	16
Cuba	14	6	11
Hungary	11	12	7

Use **MINITAB** when appropriate to help respond to the following questions.

(a) State an appropriate null hypothesis and alternative hypothesis for this information.

   H₀:

   H₁:

(b) What are the *degrees of freedom* for this test?

   Degrees of freedom = _____.

(c) Give the *equation* of the *test statistic* for this test. Define each symbol that you use in the equation.

(d) Give the values of the expected frequencies.

$e_{11} =$	$e_{12} =$	$e_{13} =$
$e_{21} =$	$e_{22} =$	$e_{23} =$
$e_{31} =$	$e_{32} =$	$e_{33} =$
$e_{41} =$	$e_{42} =$	$e_{43} =$
$e_{51} =$	$e_{52} =$	$e_{53} =$
$e_{61} =$	$e_{62} =$	$e_{63} =$

(e) What is the value of the *test statistic*?

Test statistic $\chi^2 = $ _____.

(f) What is the *p-value* for the test?

p-value = _____.

(g) Using a level of significance of $\alpha = 0.05$, what is the *critical value* for this test?

Critical value = _____.

(h) What can you *conclude* from this test?

226

2. For the table given in Problem 1, *multiply each entry by 2*.

Use **MINITAB** when appropriate to help respond to the following questions.

(a) Give the *equation* of the *test statistic* for this test. Define each symbol that you use in the equation.

(b) What are the *degrees of freedom* for this test?

Degrees of freedom = _____.

(c) Give the values of the expected frequencies.

e_{11} =	e_{12} =	e_{13} =
e_{21} =	e_{22} =	e_{23} =
e_{31} =	e_{32} =	e_{33} =
e_{41} =	e_{42} =	e_{43} =
e_{51} =	e_{52} =	e_{53} =
e_{61} =	e_{62} =	e_{63} =

(d) What is the value of the *test statistic*?

Test statistic χ^2 = _____.

(e) What is the *p-value* for the test?

p-value = _____.

(f) Using a level of significance of $\alpha = 0.05$, what is the *critical value* for this test?

Critical Value = _____.

(g) What can you *conclude* for this test?

(h) How do these results compare with those in Problem 1? i.e., what is the effect of doubling each entry?

(i) Can you make a general statement with regard to the value of the test statistic and the expected frequencies for such chi-square goodness-of-fit test when the observed frequencies are multiplied by the same positive integer?

3. *Add 100* to each value of the observed frequencies in Problem 1.

   Use **MINITAB** when appropriate to help respond to the following questions.

   (a) Give the values of the expected frequencies.

$e_{11} =$	$e_{12} =$	$e_{13} =$
$e_{21} =$	$e_{22} =$	$e_{23} =$
$e_{31} =$	$e_{32} =$	$e_{33} =$
$e_{41} =$	$e_{42} =$	$e_{43} =$
$e_{51} =$	$e_{52} =$	$e_{53} =$
$e_{61} =$	$e_{62} =$	$e_{63} =$

   (b) What is the value of the *test statistic*?

   Test statistic $\chi^2 =$ _____.

   (c) What is the *p-value* for the test?

   p-value = _____.

   (d) What can you *conclude* for this test?

(e) How do these results compare with those in Problem 1? i.e., what is the effect of adding 100 to each entry?

(g) Can you make any general statement with regard to the chi-square goodness-of-fit test when the same constant is added to the observed frequencies in a contingency table?

4. Interchange the second and third columns in Problem 1.

   Use **MINITAB** when appropriate to help respond to the following questions.

   (a) Give the values of the expected frequencies.

$e_{11} =$	$e_{12} =$	$e_{13} =$
$e_{21} =$	$e_{22} =$	$e_{23} =$
$e_{31} =$	$e_{32} =$	$e_{33} =$
$e_{41} =$	$e_{42} =$	$e_{43} =$
$e_{51} =$	$e_{52} =$	$e_{53} =$
$e_{61} =$	$e_{62} =$	$e_{63} =$

   (b) What is the value of the *test statistic*?

   Test statistic $\chi^2 =$ _____.

(c) What is the *p-value* for the test?

    p-value = _____.

(d) What can you *conclude* for this test? Be elaborate in your explanation.

(e) How do these results compare with the results in Problem 1? i.e., what is the effect of interchanging these two columns?

(f) Can you make any general statement(s) with regard to the chi-square goodness-of-fit test when the columns are interchanged in a contingency table?

5. An automobile manufacturer wishes to test for consumer preferences among five new models. A random sample of 500 consumers were selected. The table below shows the frequency distribution for the preferences of these customers. The objective is to determine whether the distribution of the consumers is **uniform**. *Note: Uniform implies that the probabilities for each type of vehicle are equal.*

Preferred Model	A	B	C	D	E
Frequency	225	185	230	187	173

Use **MINITAB** when appropriate to help respond to the following questions.

(a) State an appropriate null hypothesis and the alternative hypothesis for this information. Be *precise*!

H_0:

H_1:

(b) What are the **degrees of freedom** for this test?

Degrees of freedom = _____.

(c) Give the **equation** of the **test statistic** for this test. Define each symbol that you use in the equation.

(d) What are the values of the expected frequencies?

$e_1 = $ _____ $e_2 = $ _____ $e_3 = $ _____

$e_4 = $ _____ $e_5 = $ _____

232

(e) What is the value of the *test statistic*?

    Test statistic $\chi^2 =$ _____ .

(f) What is the *p-value* for the test?

    p-value = _____ .

(g) Using a level of significance of $\alpha = 0.05$, what is the *critical value* for this test?

    Critical Value = _____ .

(h) What can you *conclude* for this test? Be elaborate in your explanation.

6. For the table given in Problem 5, *multiply each observed frequency by 3*.

    Use **MINITAB** when appropriate to help respond to the following questions.

    (a) Give the *equation* of the *test statistic* for this test. Define each symbol that you use in the equation.

(b) What are the *degrees of freedom* for this test?

Degrees of freedom = _____.

(c) What are the values of the expected frequencies?

e_1 = _____ e_2 = _____ e_3 = _____

e_4 = _____ e_5 = _____

(d) What is the value of the *test statistic*?

Test statistic χ^2 = _____.

(e) What is the *p-value* for the test?

p-value = _____.

(f) Using a level of significance of $\alpha = 0.05$, what is the *critical value* for this test?

Critical value = _____.

(g) What can you *conclude* for this test?

(h) How do these results compare with Problem 5? i.e., what is the effect of multiplying each observed frequency by 3?

(i) Can you make any general statements with regard to the values of the expected frequencies and test statistic for such chi-square goodness-of-fit test when the observed frequencies are multiplied by the same positive integer?

7. In Problem 5, rearrange the observed frequencies as follows: 230 (C), 187 (D), 173 (E), 225 (A), and 185 (B).

(a) What are the values of the expected frequencies?

$e_1 = $ _____ $e_2 = $ _____ $e_3 = $ _____

$e_4 = $ _____ $e_5 = $ _____

(b) What is the value of the *test statistic*?

Test statistic $\chi^2 = $ _____.

(c) Using a level of significance of $\alpha = 0.05$, what is the *critical value* for this test?

Critical value = _____.

(d) What is the *p-value* for the test?

p-value = _____.

(e) What can you *conclude* for this test?

(f) How do these results compare with Problem 5?

(g) What general statement(s) can be made if the categories in a multinomial experiment are rearranged when using the chi-square goodness-of-fit test.

8. Change the observed frequency of 225 to 1000 in Problem 5.

Use **MINITAB** when appropriate to help respond to the following questions.

(a) What is the value of the *test statistic*?

Test statistic $\chi^2 =$ _____.

(b) What is the *p-value* for the test?

   p-value = _____.

(c) What can you *conclude* for this test?

(d) How do these results compare with Problem 5? Describe the effect of the extreme observed frequency.

9. **Team Exploration Project**. Use your library or any other resource to collect a set of data that could be categorize as a multinomial experiment. Use **MINITAB** to do a complete analysis of the data. Present a report of your analysis. If you select data from published research, you should give it as a reference and include a short discussion for the motivation of the study. *Note:* You will need information relating to theoretical and observed proportions.

10. **Team Exploration Project.** Use your library or any other resource to collect a set data that could be classified according to two characteristics. Use **MINITAB** to do a complete analysis of the data (test for independence or homogeneity). Present a report of your analysis. If you select data from published research, you should give it as a reference and include a short discussion for the motivation of the study.

NOTES

STATISTICS LAB #12

QUALITY *or* PROCESS CONTROL

PURPOSE - to use MINITAB to

1. construct **RUN** - *charts*
2. construct **R** - *charts*
3. construct $\overline{\text{X}}$- *charts*
4. construct **P** - *charts*

BACKGROUND INFORMATION

1. **Process data** are data that are observed over some sequence of time.

2. A **RUN**-*chart* is a sequential plot of the observed values over time.

3. A process is within *statistical control* if there are only natural variations within the process as displayed in its run chart.

4. *Random variations* are variations within the process that are due to chance.

5. *Assignable variations* are variations within the process whose causes can be identified.

6. A *control chart* of a process consists of data values plotted over time. Such a chart includes a *center line*, a *lower control limit (LCL)*, and an *upper control limit (UCL)*.

7. **The R-***chart* - These are control charts that are used to monitor variation.

8. Let $\overline{R}$ be the mean of the sample ranges. The basic format for the R-*chart* will include the sample (subgroup) ranges plotted in sequence; a center line ($\overline{R}$); the upper control limit $D_4\overline{R}$; and the lower control limit $D_3\overline{R}$. The values for D_3 and D_4 are tabulated values and are usually given in any text that deals with R-*charts*.

9. **The $\overline{\text{X}}$-***chart* - These are control charts that are used to monitor means.

10. Various approaches can be used to determine the locations of the control limits for the $\overline{\text{X}}$-*chart*. In this *Lab*, we will use the pooled standard deviation. Since the equations are rather complex, we will not list them here.

11. **The P-*chart*** - These are control charts that are used to monitor the proportion p (for example, the proportion of defective items, etc.)

12. Let $\bar{p}$ be the pooled estimate of the proportion of the characteristic (defective items) in a process. Let $\bar{q} = 1 - \bar{p}$ and n be the sample size. The basic format for the $\bar{p}$-*chart* will include the sample (subgroup) proportions plotted in sequence; a center line ($\bar{p}$); the upper control limit $\bar{p} + 3\sqrt{\dfrac{\bar{p}\bar{q}}{n}}$; and the lower control limit $\bar{p} - 3\sqrt{\dfrac{\bar{p}\bar{q}}{n}}$.

13. ***General criteria to determine whether a process is not statistically stable or out of statistical control:***

   (a) The control chart shows a pattern, a trend, or a cycle that is not random.

   (b) There is at least one point outside the upper or lower control limits.

   (c) There are eight consecutive points all above or all below the center line. This is sometimes called the ***Run of Eight Rule***.

PROCEDURES

First, load the **MINITAB** (windows version) software as in *Lab #0*.

1. RUN CHARTS

Example 1: Catch the Wave Bottling Company produces 12-ounce bottles of its special tropical punch. Twelve bottles were sampled over a 30 minute period from the bottling process. The amount (fluid ounces) of punch in each bottle was measured with the data given below. Use **MINITAB** to construct a **RUN-*chart (Individual chart)*** for this information.

1	*2*	*3*	*4*	*5*	*6*	*7*	*8*	*9*	*10*	*11*	*12*
10.2	11.6	11.4	13.2	12.7	11.3	12.6	10.8	13.3	12.4	11.9	12.1

Enter the values in C1 and rename as *WEIGHT*. Observe that the smallest value is 10.2 and the largest value is 13.3. Select **Stat→Control Charts→Individuals** and the **Individuals Chart** dialog box will appear. Select the variable *WEIGHT* for the **Variable** text box. This dialog box is shown in figure *Minitab 12.1*. Since no historical mean or standard deviation were given, **MINITAB** will use the sample data to estimate **mu** and **sigma**. Observe also that under **Tests For Special Causes**, the option **None** was selected since all we need is a **RUN-***chart* for the data. You can select the different options (**Annotation, Frame, Regions**, etc.) in the dialog box for a title, etc. on your ***run chart***. *Note:* When the **Individuals Chart** is displayed, it includes a center line (the mean $\bar{\bar{x}}$ of all the values), and three standard-deviation control limits ($\bar{\bar{x}} \pm 3s$) where s is the sample standard deviation.

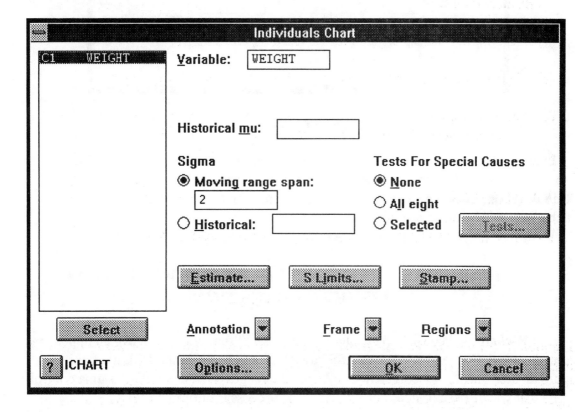

Minitab 12.1

The **RUN-***chart* for this data is shown in figure *Minitab 12.2*. Observe that the process during the given 30 minutes was rather stable about the mean value of 11.96 when you apply **BACKGROUND INFORMATION #7**.

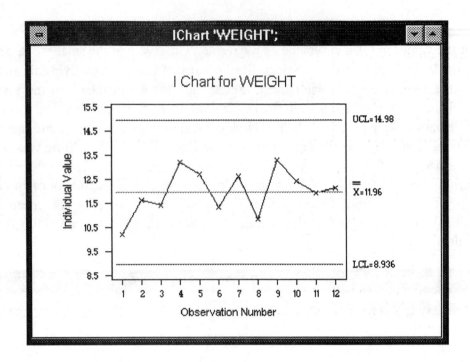

Minitab 12.2

In developing control charts for monitoring variations in a process, it would be ideal to use some measure of variation such as the standard deviation or the range. Since the **R-*chart*** (**range chart**) historically has been used to monitor the variation in a process, such charts will be discussed in the next section.

2. THE R-CHART *(CONTROL CHART FOR MONITORING PROCESS VARIATION)*

The **R-*chart*** will consist of a center line $\overline{R}$ (mean of the sample ranges), an upper control limit (UCL), and a lower control limit (LCL). The UCL and LCL are approximately the 99.7% confidence limits for the ranges. These are the 3-sigma control limits that will be used to determine whether the process is out of control.

Example 2: A new machine is installed to manufacture ball bearings whose target weight is 6 ounces. The machine is monitored every 20 minutes and a bearing is sampled and weighed. This sampling process is continued for 13 consecutive hours. The three observed weights per hour for the bearings are given below. Use **MINITAB** to construct an **R-*chart*** for this process to determine whether the process is out of control.

HOUR

1	2	3	4	5	6	7	8	9	10	11	12	13
5.98	6.01	5.97	5.98	5.97	5.99	6.01	5.99	5.97	6.03	5.98	5.99	5.98
6.01	5.99	5.99	5.97	5.98	5.98	5.97	6.00	6.02	6.01	5.99	5.98	6.00
5.99	6.02	6.01	6.01	6.00	6.03	5.98	5.98	5.98	5.98	5.97	6.03	5.99

First, enter the values in C1 and rename as *BALL*. ***Note:*** you must enter the three values for the first hour, then the three values for the second hour, and so on, i.e. , enter the values in the sequence 5.98, 6.01, 5.99, 6.01, 5.99, 6.02, 5.97, etc. This is needed in order to have samples (subgroups) of size three. Next select **Stat→Control Charts→R** and the **R-*chart*** dialog box will appear. Select **BALL** for the **Variable** box and select **Subgroup size** and type in *3,* i.e., the sample size will be three and **MINITAB** will select the first three values as the first sample, the second three values as the second sample, etc. The **R Chart** dialog box is shown in figure *Minitab 12.3*. In this example, the **Rbar estimate** is used for σ since the subgroup size n is such that $2 \leq n \leq 5$. However, when the sample size n is larger we should use other charts such as the **S** chart (not discussed in this *Lab*).

Minitab 12.3

243

Select the **OK** button and the **R-*chart*** will be displayed. For the display in figure *Minitab 12.4*, 10 major ticks are used for the vertical axis and 13 are used for the horizontal axis. Based on the criteria for an out of statistical control process in the section on **BACKGROUND INFORMATION #13**, we can safely say that the process is ***not*** out of control, i.e., the process is statistically stable. There is not a pattern of increasing or decreasing subgroup ranges.

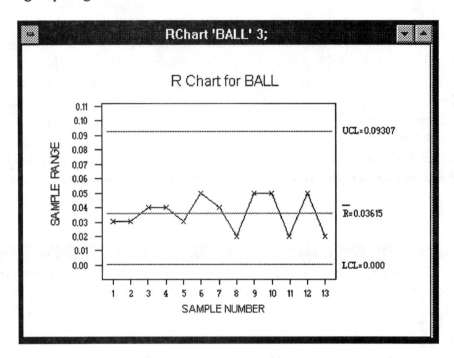

Minitab 12.4

Observe that the heading of the output in figure *Minitab 12.4* specifies the variable *(BALL)* and the sample (subgroup) size of 3.

3. THE $\overline{\text{X}}$-CONTROL CHART

Example 3: Use **MINITAB** to construct the $\overline{\text{X}}$- ***control chart*** for the data in ***Example 2*** using a sample (subgroup) size of 3.

Enter the values in column C1 *(BALL)*. Next, select **Stat→Control Charts→Xbar** and the **Xbar Chart** dialog box will be displayed. In the **Variable text** box select C1. Select the **Subgroup size** check box and type *3* in the text box. We use a value of ***three*** here because the sample (subgroup) size is equal to three. You can click on **Annotation**, **Frame**, etc. to enter a title etc. as you did in previous ***Lab*** sessions. The **Xbar Chart** dialog box with the appropriate entries is shown in figure *Minitab 12.5*.

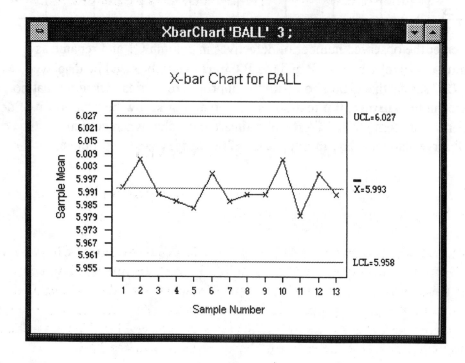

Minitab 12.5

Click on the **OK** button and the $\overline{X}$ - *chart* will be displayed in the **Xbar Chart** window. This is shown in figure *Minitab 12.6*. Observe from this chart that the process is working properly since the Xbar values are within the control limits, i.e., the sample means are within the values UCL (upper control limit) = 6.027 and LCL (lower control limit) = 5.958. Also, observe that there is no apparent pattern.

Minitab 12.6

Note: Some practitioners prefer to base the center line and control limits for the $\overline{X}$-*chart* on the sample ranges. To use the sample ranges, you could have selected the **Rbar estimate** for **Sigma** in figure *Minitab 12.5*.

4. CONTROL CHARTS FOR FRACTION DEFECTIVE OR P-CHARTS

This section of the *Lab* deals with control charts for an attribute, such as the proportion *p* of defective items. The objective is to monitor a process to determine whether the quality of a product is being maintained.

Example 4: A manufacturing company produces rear-view mirrors for a particular sports vehicle. Each day 200 mirrors are randomly selected and checked for defects. The results are given below for 20 consecutive days. Use **MINITAB** to help construct a control chart for *p* (proportion of defectives) and then interpret the chart.

Day	1	2	3	4	5	6	7	8	9	10
Number of Defectives	6	12	4	0	2	8	6	17	2	4

Day	11	12	13	14	15	16	17	18	19	20
Number of Defectives	10	10	0	6	8	2	12	6	4	8

First, enter the observed number of defectives in column C1 and rename as *DEF*. Select **Stat→Control Charts→P** and the **P Chart** dialog box will be displayed. Figure *Minitab 12.7* shows this dialog box with the appropriate entries. Observe that no value was entered in the **Historical p** text box since we do not know the proportion of defective mirrors for the process. **MINITAB** will estimate this value when computing the control limits. Observe also that the subgroup size is 200, as given in the problem.

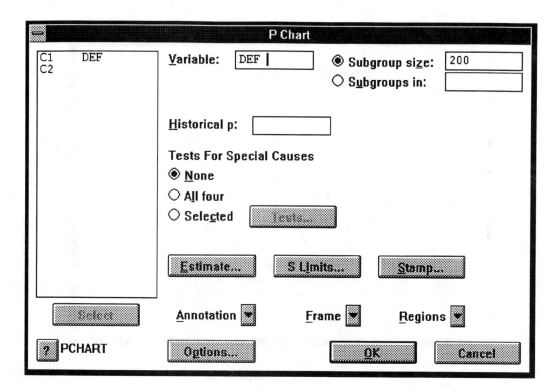

Minitab 12.7

Click on the **OK** button and the chart will be drawn as in figure *Minitab 12.8*. The **Annotation**, etc. options were used to present figure *Minitab 12.8*. Twelve ticks were chosen along the *SAMPLE NUMBER* axis using the **Frame** option in figure *Minitab 12.7*. Based on our out-of-control criteria we can say that the process is out of statistical control. However, since only one point is outside the control limits, this might be an extremely rare case or an indication that the manufacturing process is not functioning as we might expect.

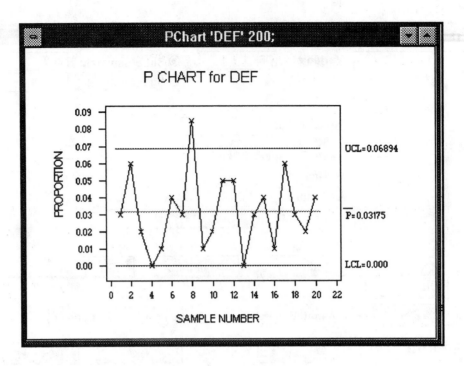

Minitab 12.8

LAB #12: *DATA SHEET*

Name: _____   *Date:* _____

Course #: _____   *Instructor:* _____

1. The following eight sets of data were observed for eight different processes. Construct **RUN-*charts*** for these data sets. Present a report with these charts and discuss their patterns.

Data #1	Data #2	Data #3	Data #4	Data #5	Data $ 6	Data #7	Data #8
13	15.1	20.1	9.4	21	20.4	10	20
13.3	15.3	19.3	9.3	20.3	20.7	10.3	20.2
12.7	14.8	20	9.7	20.9	19	9.8	20.4
12.5	14.5	20.5	9.5	17.1	21.9	9.5	20.6
13.4	15.5	19.1	9.7	17.3	21.3	9.7	20.4
13.1	15.2	19.3	9.6	19.3	22.8	10.4	20.1
13	14.9	19.5	9.8	17.2	23.2	9.3	19.9
12.8	14.8	20	10.1	17.3	24	10.9	19.6
12.6	14.7	19.3	9.7	16.8	24.2	8.9	19.2
12.9	15.9	13.7	15.3	17.4	24.4	11.4	19.5
12.7	19.9	14.1	16	17.2	23.8	8.3	19.9
9.9	16	13.8	15.5	17	24.3	11.9	20.2
12.9	15.2	13.6	15.3	16.8	24.4	8.5	20.6
13.3	15.6	13.7	15.1	17.3	25.5	12.4	20.7
13.5	14.7	13.5	16.5	16.9	25.3	8	20.5
13	14.4	13.7	16	16.6	24.9	12.8	20.1
13.1	15	13.3	15.3	17	26.1	7.6	19.9
12.8	14.9	13.4	15.9	16.7	27.9	13.1	19.6

2. A toothpaste manufacturing company recently started manufacturing an 8 ounce size baking soda toothpaste. For 26 consecutive 20-minute period of manufacturing, a sample of four per 20 minutes was taken from the process and weighed. The data (in ounces) are presented below. Construct a **R-*chart*** to determine whether the manufacturing process is out of control. Discuss your results in a report.

Note: #1, #2, etc. represent the sample number. You should read the table as follows - for sample #1, the observed values are 7.95, 7.92, 8.07, and 7.79; for sample #2, the observed values are 8.04, 7.81, 7.99, 7.96, etc.

#1	#2	#3	#4	#5	#6	#7	#8	#9	#10	#11	#12	#13
7.95	8.04	8.03	8.02	8.01	8.04	7.90	8.11	8.03	8.05	8.00	8.06	8.08
7.92	7.81	7.97	7.96	8.02	7.96	8.04	8.23	7.81	7.88	7.82	7.95	8.05
8.07	7.99	8.11	8.09	7.93	8.02	8.09	8.14	8.41	8.08	8.07	8.13	8.01
7.79	7.96	8.14	8.03	7.73	7.94	8.01	7.99	7.89	7.86	8.10	7.81	7.68

#14	#15	#16	#17	#18	#19	#20	#21	#22	#23	#24	#25	#26
8.12	7.97	7.89	8.05	8.12	7.65	7.09	7.89	8.07	7.89	7.91	8.31	8.01
7.47	7.91	8.01	8.01	8.13	7.79	7.11	8.02	8.38	7.92	7.87	8.01	8.12
8.06	8.15	8.07	8.09	8.17	7.21	7.07	7.93	8.18	7.79	8.07	7.63	7.89
8.03	7.77	8.03	8.11	8.21	7.25	7.81	7.82	8.13	8.03	8.01	7.91	8.25

3. Use **MINITAB** and the data in Problem 2 to construct an $\overline{X}$-chart. Use both estimates **Pooled Std.dev.** and **Rbar estimate** to estimate the population standard deviation (see figure *Minitab 12.5*). Discuss your results in a report.

4. The Techcell Company manufactures portable telephones. Each day 250 telephones are randomly selected and tested. The number of defective telephones per 250 for 30 consecutive days are given below. Use **MINITAB** to construct a **P-chart** for monitoring the proportion of defectives. Discuss your results.

Day 1	Day 2	Day 3	Day 4	Day 5	Day 6	Day 7	Day 8	Day 9	Day 10
10	8	6	11	4	5	12	11	8	9

Day 11	Day 12	Day 13	Day 14	Day 15	Day 16	Day 17	Day 18	Day 19	Day 20
7	4	6	14	7	6	3	9	17	6

Day 21	Day 22	Day 23	Day 24	Day 25	Day 26	Day 27	Day 28	Day 29	Day 30
10	12	11	9	4	5	8	10	7	11

5. **Team Exploration Project**. Use your library or any other resource to collect a set of data that could be analyzed with the **R-chart**. Use **MINITAB** to analyze the data. Present any relevant analysis and discussions in a report.

6. **Team Exploration Project**. Use your library or any other resource to collect a set of data that could be analyzed with the $\overline{X}$-*chart*. Use **MINITAB** to analyze the data. Present any relevant analysis and discussions in a report.

7. **Team Exploration Project**. Use your library or any other resource to collect a set of data that could be analyzed with the **P-chart**. Use **MINITAB** to analyze the data. Present any relevant analysis and discussions in a report.

8. **Team Exploration Project**. Use **MINITAB** to simulate 40 subgroups of size 25 integer values between 0 and 9. Let the number of odd numbers in each of the 40 sets represent the number of defectives of a process. Use the theory of this chapter to determine whether the process is out of control. Present your findings in a report. *(Note: Each set of values generated is equivalent to a binomial experiment with n= 25, p = 1/2.)*

9. **Team Exploration Project**. Use **MINITAB** to simulate 40 subgroups of size 20 integer values between 0 and 9. Let the number of zeros in each of the 40 sets represent the number of defectives of a process. Use the theory of this chapter to determine whether the process is out of control. Present your findings in a report. *(Note: Each set of values generated is equivalent to a binomial experiment with n= 20, p = 1/10.)*

NOTES

STATISTICS LAB # 13

ONE-WAY ANALYSIS OF VARIANCE

PURPOSE - to use MINITAB to

1. perform a *One-Way* or *One-factor* or *Single-Factor Analysis of Variance*

BACKGROUND INFORMATION

1. **Analysis of Variance (ANOVA)** is a method of testing the equality of three or more population means by analyzing sample variances. (See general notation in item 9).

2. The **ANOVA** method uses the *F*-distribution.

3. Assumptions applied when testing the hypothesis that three or more samples come from populations with the same mean:

   (a) The populations have normal distributions.

   (b) The populations have the same variance (or standard deviation).

   (c) The samples are random and independent of each other.

4. The requirements of normality and equal variances can be relaxed somewhat unless a population has a distribution that is very non-normal or the population variances differ by large amounts.

5. If the sample sizes are equal or nearly equal, one variance can be up to nine times larger than another variance and the results of the **ANOVA** will still be reliable.

6. **ANOVA** is based on a comparison of two different estimates of the variance common to the different populations: (a) *variance between samples* (b) *variance within samples.*

7. The term *one-way analysis of variance* (*one-* or *single-factor analysis of variance*) is used because the sample data are separated into groups according to *one characteristic* or *factor*.

8. A ***treatment*** (or *level*) is a property, or characteristic, that enables us to distinguish the different populations from each other.

9. ***Notation:*** k = number of population means (number of samples); n_i = number of values in the ith sample for i = 1, 2, ... , k;

N = combined sample size ($\sum\limits_i^k n_i$);

$\overline{\overline{x}}$ = overall mean (mean of all the values);

$\overline{x}_i$ = mean of sample i;

s_i^2 = variance of sample i.

10. ***Variance Between Samples*** $= \dfrac{\sum n_i (\overline{x}_i - \overline{\overline{x}})^2}{k-1}$.

11. ***Variance Within Samples*** $= \dfrac{\sum (n_i - 1) s_i^2}{\sum (n_i - 1)}$.

12. Test statistic $F = \dfrac{variance\ between\ samples}{variance\ within\ samples}$.

13. The null hypothesis to be tested $\mathbf{H_0}$: $\mu_1 = \mu_2 = \mu_3 = ... = \mu_k$ (the means are equal) versus the alternative $\mathbf{H_1}$: the means are not all equal.

14. The numerator degrees of freedom for the critical F value is $(k - 1)$ and the denominator degrees of freedom is $(N - k)$.

15. The ***Mean Square*** for the **Factor (treatments or levels)** in the **MINITAB** output corresponds to the ***Variance Between Samples***.

16. The ***Mean Square*** for the **Error** in the **MINITAB** output corresponds to the ***Variance Within Samples***.

PROCEDURES

Load the **MINITAB** (windows version) software as in *Lab #0*.

1. ONE-FACTOR ANALYSIS OF VARIANCE

Note: *Some elementary statistics texts will consider two cases for the one-factor ANOVA: (a) one-factor ANOVA with equal sample sizes and (b) one-factor ANOVA with unequal sample sizes. The formulas above will hold for both cases and the MINITAB procedures are the same for both cases.*

Example 1: A psychologist wants to investigate the effect of social background on the time (in minutes) it takes freshmen to solve a puzzle. A random sample of students from different backgrounds is selected, resulting in the following data[*]. Use **MINITAB** to test whether social background has no effect on the time required to solve the puzzle, i.e., test whether the average time to solve the puzzle for the different background groups is the same. Use $\alpha = 0.05$.

Inner City	*Urban*	*Suburban*	*Rural*
16.5	10.9	18.6	14.2
5.2	5.2	8.1	24.5
12.1	10.8	6.4	14.8
14.3	8.9		24.9
	16.1		5.1

First, enter the data values for the four categories in columns C1, C2, C3, and C4. Rename these columns *IC* (Inner City), *UR* (Urban), *SU* (Suburban), and *RU* (Rural). Note that this is a *one-factor analysis of variance*, the factor being *social background* and the four populations (levels) from which the samples were taken are IC, UR, SU, and RU. To use **MINITAB**, select **Stat→ANOVA→Oneway (Unstacked),** and the **Oneway Analysis of Variance** dialog box will be displayed. Select IC, UR, SU, and RU for the **Responses (in separate columns)** text box as shown in figure *Minitab 13.1*.

[*] See page 433, *Statistics - an Introduction*, 4th ed., Mason, Lind & Marchal; Saunders College Publishing

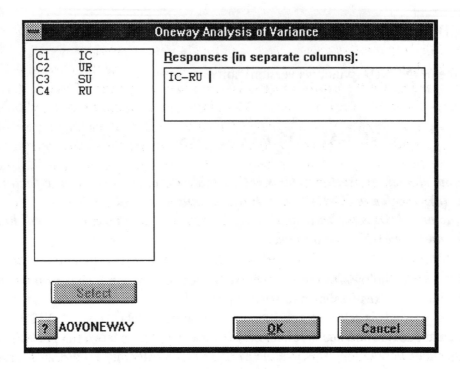

Minitab 13.1

Use the mouse to click on the **OK** button and the **Session** window will display the results of the computations for the test. The **Session** window is shown in figure *Minitab 13.2*.

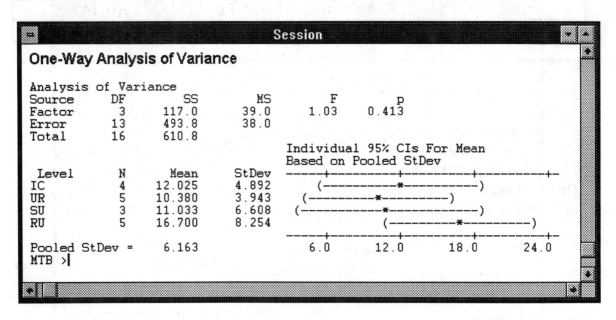

Minitab 13.2

The *degrees of freedom* for the *between samples* (**Factor**) *sum of squares* is three. This corresponds to the *numerator* degrees of freedom for the *critical F* value. The *denominator* degrees of freedom for the critical *F* value is 13 and this corresponds to the degrees of freedom for the *within samples* (**Error**) *sum of squares*. Observe that the *mean square for the between samples* is 39 and the *mean square for the within samples* is 38. These values are the sum of squares divided by the corresponding degrees of freedom. The *ratio* of the *between samples mean square* and the *within samples mean square* gives the value of the *F* test statistic of 1.03 with a corresponding *p-value* of 0.413. Also, individual 95% confidence intervals for the population means (IC, UR, SU, and RU) are displayed in the window. With this output you can now write up the hypothesis test for the example within the guidelines of your instructor.

Note: Since the p-value = 0.413 > 0.05 you should fail to reject the hypothesis of equal means and conclude that there is insufficient evidence to claim that the means are not all equal.

If you use the *classical* approach to perform the hypothesis test, then you will need the critical (table) *F* value. You can use **MINITAB** to compute this *critical F* value. Select **Calc→Probability Distribution→F** and the **F Distribution** dialog box will be displayed. Figure *Minitab 13.3* shows this dialog box with the appropriate entries when the level of significance is $\alpha = 0.05$. Observe that the **Input constant** text box has the value 0.95 so the critical *F* value will have an area of 0.05 to the right of it.

Minitab 13.3

Click on the **OK** button and the critical *F* value of 3.4105 will be displayed in the **Session** window as shown in figure *Minitab 13.4*. Again the *test statistic F* value of 1.03 is *less than* the *critical F* value of 3.4105 and hence you will *fail to reject the null hypothesis* of equal means.

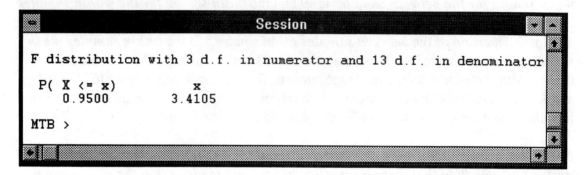

Minitab 13.4

Note: You can check for the normality assumption by constructing histograms and viewing them or use normal probability plots.

LAB #13: *DATA SHEET*

Name: _____   *Date:* _____

Course #: _____   *Instructor:* _____

1. An independent testing agency has been hired by a large manufacturer of tires. Specifically, the tire manufacturer would like to know if there is a difference in the tread wear of a tire on a variety of road surfaces. To assure uniformity, the testing agency measured the wear on only the right front tire, drove the cars in all types of traffic, and attempted to ensure that the weight was the same in all test vehicles. The data, in millimeters of wear, are shown below[*].

Type of Road Surface			
Concrete	**Composite**	**Brick**	**Gravel**
8.8	10.1	11.9	13.4
9.6	10.1	11.1	13.0
8.3	10.3	11.0	11.9
9.3	9.8	12.1	12.6
9.1	9.9	12.6	12.7
8.3	10.6	10.9	13.0
8.4	10.8	11.8	
	10.3	12.9	
		12.3	

Suppose you were to test, at the 0.05 level of significance, whether there is a difference in the mean amount of wear from the various surfaces. When appropriate, use **MINITAB** to help respond to the following questions.

(a) What is the single *factor* in this experiment? _____.

(b) What are the *treatments* (*levels*) of the factor? _____.

(c) What is the value of the *between samples sum of squares*? _____.

(d) What is the value of the *within samples sum of squares*? _____.

(e) What are the *degrees of freedom* for the between samples sum of squares? _____.

(f) What are the *degrees of freedom* for the within samples sum of squares? _____ .

(g) What is the value of the *mean square* for the between samples? _____ .

(h) What is the value of the *mean square* for the within samples? _____ .

(i) What is the value for the *test statistic*? $F =$ _____ .

(j) What is the *p-value* for the test? _____ .

(k) What is the critical value for the test? $F =$ _____ .

(l) Which treatment (level) has the largest tread wear? Analyze the confidence intervals. Discuss your observations.

(m) Write up a complete hypothesis test as presented (required) by your instructor.

2. It is suspected that the environmental temperature in which batteries are activated affects their activated lives. Thirty homogeneous batteries were tested, six at each of five temperatures, and the data is shown below[*].

Temperature (^0C)				
0	**25**	**50**	**75**	**100**
55	60	70	72	65
55	61	72	72	66
57	60	73	72	60
54	60	68	70	64
54	60	77	68	65
56	60	77	69	65

Suppose you were to test, at the 0.02 level of significance, whether there is a difference in the mean battery life for the different temperatures. When appropriate, use **MINITAB** to help respond to the following questions.

(a) What is the single *factor* in this experiment? _____.

(b) What are the *treatments* (*levels*) of the factor? _____.

(c) What is the value of the *between samples sum of squares*? _____.

(d) What is the value of the *within samples sum of squares*? _____.

(e) What are the *degrees of freedom* for the between samples sum of squares? _____.

(f) What are the *degrees of freedom* for the within samples sum of squares? _____.

(g) What is the value of the *mean square* for the between samples? _____.

(h) What is the value of the *mean square* for the within samples? _____.

(i) What is the value for the *test statistic*? $F =$ _____.

(j) What is the *p-value* for the test? _____.

(k) What is the critical value for the test? $F =$ _____.

(l) What temperature is best for the batteries to be operating? Analyze the confidence intervals. Discuss your observations.

(m) Write up a complete hypothesis test as presented (or required) by your instructor.

3. **Team Exploration Project**. Use your library or any other resources to collect data for a single-factor experiment. Use **MINITAB** to do an appropriate hypothesis test. Establish whether these samples are independent and are normally or approximately normally distributed by constructing histograms for the data values for the different treatments (levels). Write up a report with your discussions and findings. If you use data from a published report, you should include relevant discussions about the reliability of your results and the study.

4. **Team Exploration Project**. Design your own single-factor experiment and collect data based on this design. Use **MINITAB** to do an analysis of variance and present your results in a report.

♦ See page 436, *Statistics - an Introduction*, 4th ed.; by Mason, Lind & Marchal; Saunders College Publishing.

♣ See page 88, *Fundamental Concepts in the Design of Experiments*, 4th ed.; by Hicks; Saunders College Publishing.

NOTES

STATISTICS LAB # 14

TIME SERIES ANALYSIS

PURPOSE - to use MINITAB to

1. present a *time series plot*
2. perform a *trend analysis*
3. analyze *Moving Averages*
4. analyze *Exponential Smoothing*
5. perform a *classical decomposition* to model and forecast a time series

BACKGROUND INFORMATION

1. A *time series* is a set of observations measured for a variable at successive points in time or over successive periods of time.

2. A *trend* (or *secular trend*), **T**, is a long-term, relatively smooth pattern that the time series exhibits. The *trend* is not always linear.

3. A *cycle*, **C**, is a wavelike pattern over long periods of time. In general, any recurring sequence of points above or below the trend line lasting for more than one year can be associated with the cyclical component of the time series.

4. *Seasonal variation*, **S**, is like a cycle that occurs over short repetitive calendar periods, and by definition, have durations of less than one year.

5. *Random variation*, **R**, or the irregular component of the time series, accounts for the random variability in the time series.

6. *Models for Time Series:*

   (a) *Additive Model*-- the value of the time series at time t, $\mathbf{Y_t}$, is given by

   $$\mathbf{Y_t = T_t + C_t + S_t + R_t} \,.$$

   (b) *Multiplicative Model* -- the value of the time series at time t, $\mathbf{Y_t}$, is given by

   $$\mathbf{Y_t = T_t \times C_t \times S_t \times R_t} \,.$$

267

7. The *Moving Average* method uses the simple arithmetic average of the most recent *n* data values in the time series as the forecast for the next period.

8. In the moving average method, all observations in the calculation receives the same weight. In the case of the *Weighted Moving Average* a different weight can be assigned to the observations, with the sum of the weights equal to one.

9. Disadvantages associated with the *Moving Average* methods:

   (i) there are no moving averages for the first and last sets of time periods (loss of information)

   (ii) the moving average is "memoryless", i.e., the moving average "forgets" most of the previous time series values.

10. *Exponential Smoothing* uses a weighted average of past time series data values as the forecast. The two disadvantages listed in #9 are not present when applying the *Exponential Smoothing* method.

11. The *Exponential Smoothing model* is given as:

$$F_{t+1} = \alpha Y_t + (1 - \alpha)F_t,$$

   where

   F_{t+1} = exponentially smoothed forecast of the time series for period $t+1$
   Y_t = actual value of the time series at time t
   F_t = exponentially smoothed forecast of the time series for period t
   α = smoothing constant, where $0 \leq \alpha \leq 1$

   Note: $F_1 = Y_1$.

12. *Linear Model (empirical) for Long Term Trend:*

$$\hat{Y}_t = b_0 + b_1 t, \text{ where } t \text{ is the time period.}$$

13. The procedure that is used to identify the *cyclical variation* is called the *percentage of trend*. The *percentage of trend* is computed as follows:

   (a) determine the trend line;

   (b) for each time period, compute the trend value $\hat{Y}_t$;

(c) the percentage of the trend is $(Y_t / \hat{Y}_t)100\%.$

14. You can use *moving averages* to remove *seasonal effect* and *random variations.*

The following steps can be used:

(a) Let the number of moving average periods be equal to number of *types* of seasons.

(b) If the number of periods is even, compute the *centered moving averages*.

(c) The moving averages reduce $Y_t = T_t \times C_t \times S_t \times R_t$ to $MA_t = T_t \times C_t$, i.e., the moving average has the effect of removing S_t and R_t from the model.

(d) Compute $Y_t/(MA_t) = S_t \times R_t$. This is a measure of the *seasonal and random variations*.

(e) Remove the random variation by computing the average of the ratios in (d). The resulting average is a measure of the *seasonal differences*.

(f) The average of the ratios in step (e) is a measure of the *seasonal indices*.

PROCEDURES

Load the **MINITAB** (windows version) software as in *Lab #0*.

1. USING MINITAB TO PRESENT A TIME SERIES PLOT

Example 1: The following table gives the number of voice mails (VMs) per month left on a recording system for a period of 24 months. Use **MINITAB** to present a time series plot for this data set.

Month	Jan.	Feb.	Mar.	April	May	June	July	Aug.	Sept.	Oct.	Nov.	Dec.
VM	51	79	59	76	64	52	100	76	49	47	71	68
VM	28	66	92	37	51	79	59	75	57	54	97	72

Enter the data set in C1 and select **Graph→Time Series Plot** and the **Time Series Plot** dialog box will appear. Select C1 for **Graph1** in the **Graph variables** box. For the **Time Scale** (horizontal axis), click on **Calendar**, and in the adjacent drop down box select **Month**. You can use **Options, Annotations**, etc. to make your graph more presentable if you wish. The **Time Series Plot** dialog box is shown in figure *Minitab 14.1*.

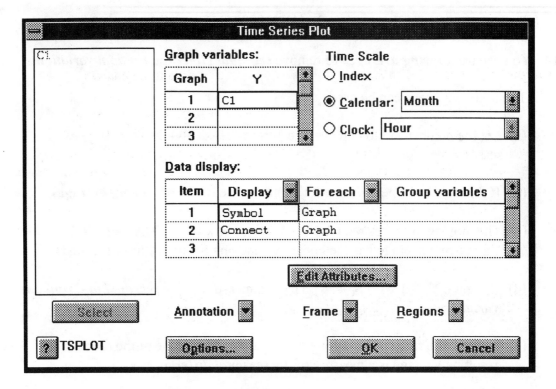

Minitab 14.1

Click on the **OK** button and the graph of the time series will be displayed in a **Graph** window. This is shown in figure *Minitab 14.2*.

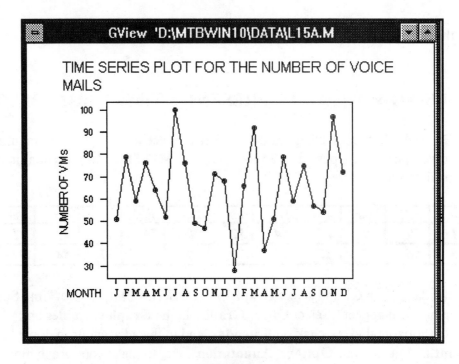

Minitab 14.2

2. TREND ANALYSIS OF A TIME SERIES

Example 2: Use the data in *Example 1* and **MINITAB** to do a trend analysis.

If the data is already entered, select **Stat→Time Series→Trend Analysis** and the **Trend Analysis** dialog box will appear. Enter the variable (or column number) in the **Variable** text box and select **Linear** for the **Model Type**. Here we are assuming that the trend is linear. You can select the **Options** button and select **Summary table and results table** to display values of the **Trend** and **Detrend** in the **Session** window. The trend is just the predicted value using the linear model and the detrend (or error) is the difference between the actual observed value and the corresponding predicted value. Select the **OK** button, along with the trend and detrend values in the **Session** window, a **Graphics** window will also be displayed as shown in figure *Minitab 14.3*.

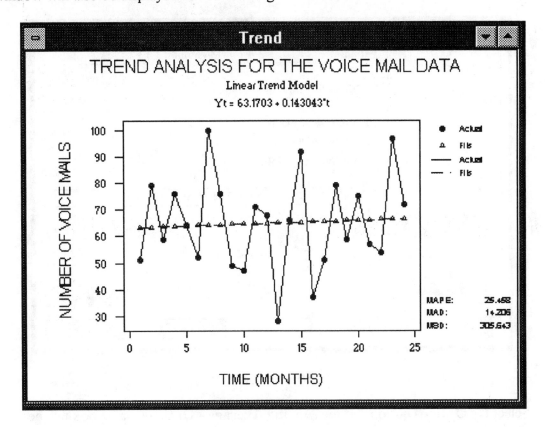

Minitab 14.3

This window displays the *equation of the trend line* ($Y_t = 63.1703 + 0.143043t$) and the graph of the line superimposed on the time series plot. Observe that the series has a slight upward trend. This graphics window also displays three measures of the accuracy of the fitted values: (a) MAPE - Mean Absolute Percentage Error (25.4580); (b) MAD - Mean Absolute Deviation (14.2058); and (c) MSD - Mean Squared Deviation (305.643).

271

These will not be discussed in this *Lab*. Note that in the **Options dialog box** you also have the option of forecasting.

3. MOVING AVERAGES OF A TIME SERIES

Example 3: Use the data in *Example 1* and **MINITAB** to do a 3-period moving average analysis for the time series.

Assume that the data is already in C1. Select **Stat→Time Series→Moving Average** and the **Moving Average** dialog box will appear. The dialog box with the appropriate entries is shown in figure *Minitab 14.4*. Observe that we are also generating forecasts for 20 periods starting at time period 3.

Moving Average	
Variable: C1	**MA length:** 3
☐ **Center the moving averages**	
☒ **Generate forecasts**	
Number of forecasts: 20	
Starting from origin: 3	
Title: 3-PERIOD MA FOR THE VOICEMAIL DATA	

Minitab 14.4

Select the **Options** button and the **Moving Average - Options** dialog box will be displayed. Here we choose **Plot smoothed vs. actual** and we selected the **Summary table and results table** option. This is shown in figure *Minitab 14.5*.

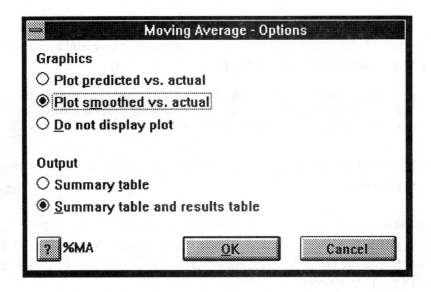

Minitab 14.5

Click on the **OK** button to take you back to the **Moving Average** dialog box
(figure *Minitab 14.4*). Select the **OK** button and the **MA** window, as well as the **Session**
window information will be generated. Figure *Minitab 14.6* shows the **MA** window.
Included in this window is a plot of the time series, a plot of the predicted (moving
average) values, and a plot of the forecast values.

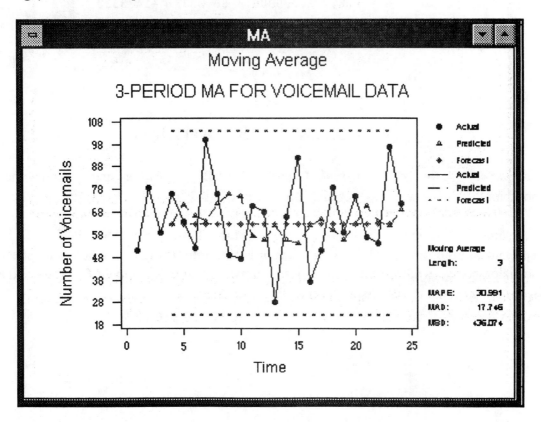

Minitab 14.6

273

Figure *Minitab 14.7* displays a portion of the **Session** window. Included in this window are the moving average values, the predicted values, etc.

```
┌─────────────────────────────────────────────────────────────┐
│ ▪                         Session                      ▾  ▴  │
├─────────────────────────────────────────────────────────────┤
│ Moving average                                            ▪  │
│                                                              │
│ Data       C1                                                │
│ Length     24.0000                                           │
│ NMissing 0                                                   │
│                                                              │
│ Moving Average                                               │
│ Length: 3                                                    │
│                                                              │
│ Accuracy Measures                                            │
│ MAPE:   30.991                                               │
│ MAD:    17.746                                               │
│ MSD:    436.074                                              │
│                                                              │
│                                                              │
│  Row   Period    C1        MA      Predict      Error        │
│                                                              │
│   1       1      51         *          *           *         │
│   2       2      79         *          *           *         │
│   3       3      59     63.0000        *           *         │
│   4       4      76     71.3333    63.0000     13.0000       │
│   5       5      64     66.3333    71.3333     -7.3333       │
│   6       6      52     64.0000    66.3333    -14.3333       │
│                                                              │
└─────────────────────────────────────────────────────────────┘
```

Minitab 14.7

4. EXPONENTIAL SMOOTHING

Example 4: Use the data in *Example 1* and **MINITAB** to apply the (single) exponential smoothing technique with a smoothing constant of $\alpha = 0.2$.

Assume that the data is already in C1. Select **Stat→Time Series→Single Exponential Smoothing** and the **Single Exponential Smoothing** dialog box will appear. The dialog box with the appropriate entries is shown in figure *Minitab 14.8*. Observe that we are using a smoothing constant of $\alpha = 0.2$ in the **Use** box.

Single Exponential Smoothing

Variable: C1

Weight to Use in Smoothing
○ Opti_m_ize
◉ _U_se   0.2

☐ _G_enerate forecasts
 _N_umber of forecasts:
 S_t_arting from origin:

_T_itle: SINGLE EXPONENTIAL SMOOTHING FOR EMAIL DATA

Select   Storage...

? %SES   Options...   OK   Cancel

Minitab 14.8

To obtain a plot of the actual and smoothed values, select **Options** and the options dialog box will appear. Select **Plot smooth vs. actual** to obtain the above-mentioned plots. To obtain the smoothed and predicted values, select **Summary table and results table** as shown in figure *Minitab 14.9*. The average of the first 6 values were used as the estimate for the initial smoothed value.

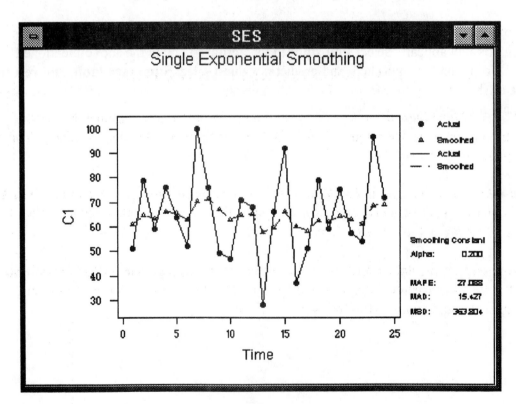

Minitab 14.9

Click on the **OK** button to take you back to the **Single Exponential Smoothing** dialog box, and click on the **OK** button. The outputs for the graph is shown in *Minitab 14.10*. The **summary** information is shown in figure *Minitab 14.11*.

Minitab 14.10

276

```
┌─────────────────────────── Session ───────────────────────────┐
│ Single Exponential Smoothing                                   │
│                                                                │
│ Data      C1                                                   │
│ Length    24.0000                                              │
│ NMissing  0                                                    │
│                                                                │
│ Smoothing Constant                                             │
│ Alpha: 0.2                                                     │
│                                                                │
│ Accuracy Measures                                              │
│ MAPE:   27.088                                                 │
│ MAD:    15.427                                                 │
│ MSD:    363.804                                                │
│                                                                │
│                                                                │
│   Row    Time     C1    Smooth    Predict     Error            │
│    1       1      51   61.0000   63.5000   -12.5000            │
│    2       2      79   64.6000   61.0000    18.0000            │
│    3       3      59   63.4800   64.6000    -5.6000            │
│    4       4      76   65.9840   63.4800    12.5200            │
│    5       5      64   65.5872   65.9840    -1.9840            │
│    6       6      52   62.8698   65.5872   -13.5872            │
│    7       7     100   70.2958   62.8698    37.1302            │
│    8       8      76   71.4367   70.2958     5.7042            │
└────────────────────────────────────────────────────────────────┘
```

Minitab 14.11

5. CLASSICAL DECOMPOSITION

Classical decomposition separates the time series into trend, seasonal, and error components by using least-squares analysis, trend analysis, and moving averages. You can also generate forecasts.

Example 5: Use the data in *Example 1* and the **Decomposition** command in **MINITAB** to obtain the trend, seasonal, and error components for a multiplicative model. Also, generate forecasts for the next 12 months.

Assume that the data is already in C1. Select **Stat→Time Series→Decomposition** and the **Decomposition** dialog box will appear. The dialog box with the appropriate entries is shown in figure *Minitab 14.12*.

Minitab 14.12

Click on the **OK** button and three **Decomp** *(graph)* windows will be generated along with the forested values in the **Session** window.

Observe that the **Session** window provides the trend line equation $Y_t = 63.1703 + 0.143043*t$ and the seasonal indices. You can use these indices to help obtain predicted and forecasted values. This portion of the **Session** window is shown in figure *Minitab 14.13*.

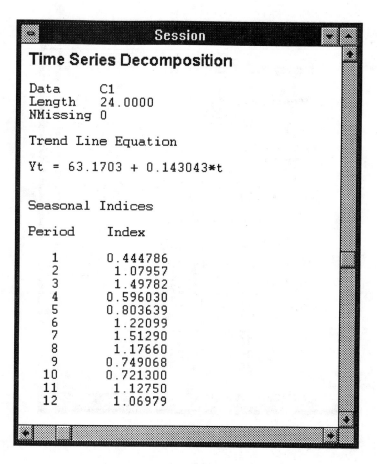

Minitab 14.13

Another portion of the **Session** window gives three measures that you can use to determine the accuracy of the fitted values: MAPE (16.57), MAD (10.422), and MSD (273.230). This portion of the **Session** window is shown in figure *Minitab 14.14*. Also included in this output are the forecasted values. Observe the two peaks at period 27 and 31.

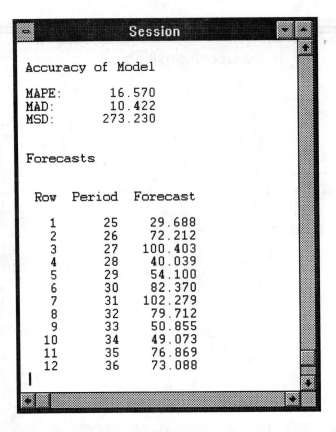

Minitab 14.14

The **Decomp1** window displays a plot of the actual, predicted, and forecasted values along with the trend line as shown in figure *Minitab 14.5*. The graph reveals a slight upward trend, with a 24-month seasonal component. The graph also reveals the divergence of the predicted values from the actual values. Included in this graph are the forecasted values for the next 12 months. The portion of the graph to the far right represents these values.

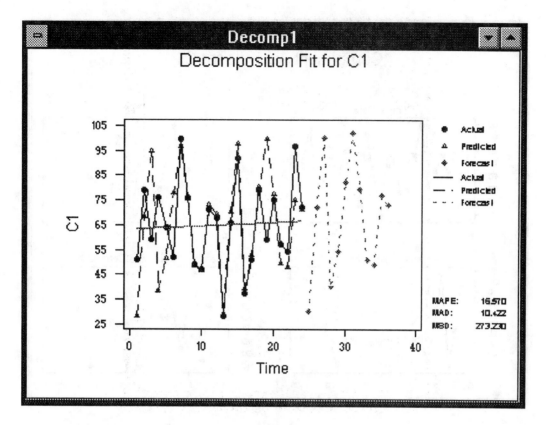

Minitab 14.15

The **Decomp2** window, as shown in figure *Minitab 14.16*, provides a component analysis for the voice-mail data. This window includes plots of the original data, detrended data, seasonally adjusted data, and the seasonally adjusted and detrended data.

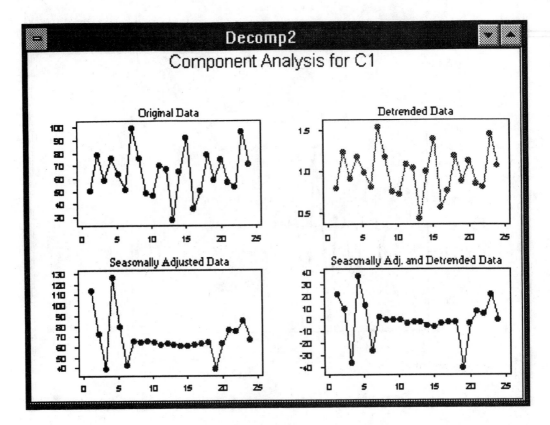

Minitab 14.16

The **Decomp3** window, as shown in figure *Minitab 14.17*, provides a seasonal analysis for the voice-mail data. It contains a plot of the seasonal indices. It also contains plots of the original data, present variation, and residuals, all by seasonal periods. Note the large negative residuals present in the residuals plots and also the one large positive residual. This could indicate that a problem may exist.

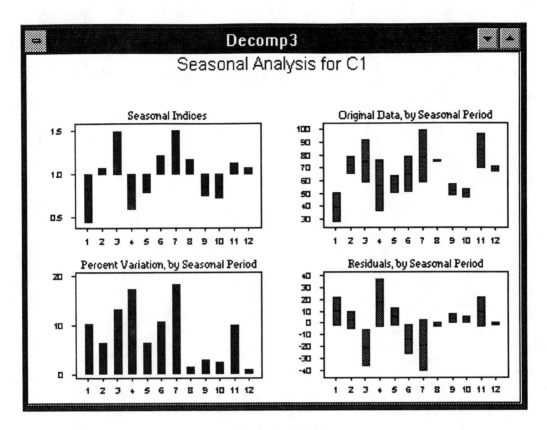

Minitab 14.17

NOTES

LAB #14: *DATA SHEET*

Name: _____   *Date:* _____

Course #: _____   *Instructor:* _____

1. **Team Exploratory Project**. Given the following 24-month time series for the amount of pieces of junk-mail (**JM**) received in a particular office:

Month	Jan.	Feb.	Mar.	April	May	June	July	Aug.	Sept.	Oct.	Nov.	Dec.
JM	33	41	48	43	54	52	59	48	49	47	71	68
JM	28	43	51	46	57	55	62	51	52	50	74	72

Use **MINITAB** to-

(a) Compute the three-period moving averages.
(b) Graph the time series and the three-period moving averages on the same plot.
(c) Does there appear to be a seasonal (yearly) pattern? Explain.
(d) Apply (single) exponential smoothing with $\alpha = 0.2$, 0.5, and 0.8.
(e) Graph the time series and superimpose the graphs for the three sets of exponentially smoothed values.
(f) Does there appear to be a trend component in the time series? Explain.
(g) Fit a linear trend line to the time series.
(h) Use this trend line to forecast the amount of junk mail for the next 24 month period.
(i) Present a plot of the time series, the trend line, and the forecasted values.
(j) Calculate the percentage of trend for each time period.
(k) Plot the percentage of the trend.
(l) Does there appear to be a cyclical effect? Explain.
(m) Calculate the monthly indices.
(n) Do a classical decomposition and discuss your results.

Present your results for (a) through (n) in the above problem in a *REPORT*.

2. **Team Exploratory Project**. Use your library or any other resource to collect a large set (at least 60 values) of time series data and repeat Problem 1.

NOTES

STATISTICS LAB #15

NONPARAMETRIC HYPOTHESIS TESTS

PURPOSE - to use MINITAB to

1. help perform the *Sign Test*
2. help perform the *Signed-Rank Test* (Wilcoxon)
3. help perform the *Rank Sum Test for Comparing Two Populations* (Mann-Whitney)
4. help perform the *Runs Test for Randomness*

BACKGROUND INFORMATION

1. A statistical test that makes no assumptions about the form of the underlying probability distribution is called a *nonparametric test*.

2. The *One -Sample Sign Test* is used to test hypotheses concerning the median of a single population.

3. The term "sign" is used for the *Sign Test* because data are converted to a series of + (plus) signs and - (minus) signs in order to do the analysis. These signs are obtained by taking the difference between the observed values and the value of the median postulated in the null hypothesis. Any zero difference is discarded and the sample size *n* is adjusted.

4. The *test statistic* for the *Sign Test*:

   (a) two-tailed test -- Minimum of (the number of + signs and the number of - signs)
   (b) right-tailed test -- number of - (minus) signs
   (c) left-tailed test -- number of + (plus) signs

5. The *decision rule* for the *Sign Test*:

   Let *k* be the value of the test statistic as described in 4. It is assumed that the distribution of *K* is binomial with parameters *n* and $p = 0.5$.

   (a) two-tailed test -- for a given significance level α, reject the null hypothesis if

   $$P\{K \le k| n, p = 0.5\} \le \frac{\alpha}{2}.$$

(b) right-tailed or left-tailed test -- for a given significance level α, reject the null hypothesis if $P\{K \le k|\ n, p = 0.5\} \le \alpha$.

(c) Decisions can also be made from a computed *p*-value as was done in previous *Labs*.

6. The *Wilcoxon Signed-Rank* test uses more information than the *Sign Test* to perform a test for the population median. It uses the signs of the differences and the ranks of the absolute values of the differences between the observed values and the value of the median postulated in the null hypothesis. Any zero difference is discarded and the sample size *n* is adjusted.

7. Let T_+ = sum of the ranks with positive signs.

Let T_- = sum of the ranks with negative signs.

The *test statistic* for the *Wilcoxon Sign Test*:

(a) two-tailed test -- Minimum of(T_+ and T_-)
(b) right-tailed test -- T_-
(c) left-tailed test -- T_+

8. The *decision rule* for the *Wilcoxon Signed-Rank Test*:

Let *k* be the value of the test statistic and assume that the distribution of *K* is binomial with parameters *n* and $p = 0.5$.

(a) two-tailed test -- for a given significance level α, reject the null hypothesis if

Minimum of(T_+ and T_-) $\le$ critical table value for *n* and selected $\dfrac{\alpha}{2}$. These are

special tables designed for this test and should be found at the end of many elementary statistics text that deals with nonparametric statistics.

(b) right-tailed -- for a given significance level α, reject the null hypothesis if $T_-\le$ critical table value for *n* and selected α.

(c) left-tailed test -- for a given significance level α, reject the null hypothesis if $T_+\le$ critical table value for *n* and selected α.

(d) Decisions can also be made from a computed *p*-value as was done in previous *Labs*.

9. The *Rank Sum (or the Mann-Whitney) Test for Comparing Two Populations* is a nonparametric test *used to compare two populations medians*.

10. The test statistic for the **Rank Sum** test is obtained by combining the two independent samples and then rank all sample values from smallest to largest. Tied observations are assigned the mean of the rank of the positions they would have occupied had there been no ties. The value of the test statistic is computed from $S - \dfrac{n_1(n_1+1)}{2}$, where S is the sum of the ranks assigned to the sample observations from **Population 1** and n_1 is the sample size from **Population 1**. Here, **Population 1** is chosen arbitrarily.

11. The critical values for the **Rank Sum** test are usually given in special tables at the end of the text. The test statistic value is compared with these critical values for the given significance level α to determine whether to reject the postulated null hypothesis.

12. A **run** is defined as a sequence of like events, items, or symbols that is preceded and followed by an event, item or symbol of a different type, or by none at all.

13. The number of events, items, or symbols in a run is referred to as its **length**.

14. The *(one-sample) Runs Test* helps us decide whether a sequence of events, items, or symbols is the result of a random process.

15. The test statistic for the **Runs Test** is the value of the number of runs.

16. The critical values for the **Runs Test** are usually given in special tables at the end of the text. The test statistic value is compared with these critical values for the given significance level α to determine whether to reject the postulated null hypothesis.

PROCEDURES

First, load the **MINITAB** (windows version) software as in *Lab #0*.

1. THE ONE-SAMPLE SIGN TEST

Example 1: A student prefers a particular brand of corn chips that is packaged with a label weight of 16 ounces. However, he believes that the bags tend to be underweight. In an attempt to prove this belief, 16 bags were purchased over a period of several weeks. The resulting weights (ounces) are given below.

16.1	15.2	14.7	15.2	15.3	15.3	15.1	15.9
15.5	15.7	15.2	16.1	15.3	15.0	15.2	15.3

Use the sign test to determine if the data present sufficient evidence, at the 0.05 level of significance, to support the student's belief.

MINITAB will be used to help provide an answer. Here, we need to test the null hypothesis that the median weight is at least 16 ounces against an alternative hypothesis that the median weight is less than 16 ounces.

In order to use the *sign test* in **MINITAB**, first enter the data values in column *C1*. Now to apply the *sign test* to column C1, select **Stat→Nonparametrics→ 1-Sample Sign** and the **1-Sample Sign** dialog box will be displayed. Select the **Test median** check box and type *16* in the text box. Select **less than** in the **Alternative** drop down box. That is, we are testing whether the median weight is at least 16 (ounces) against the alternative that it is less than 16 (ounces). The **1-Sample Sign** dialog box is shown in figure *Minitab 15.1* with the appropriate entries.

Minitab 15.1

Click on the **OK** button and the **Session window** as shown in figure *Minitab 15.2* will display the results. Since the p-value for the test is 0.0021 (rather small compared to a significance level of 0.05) we can reject the null hypothesis and claim that the median weight for the bags of corn chips is less than 16 ounces.

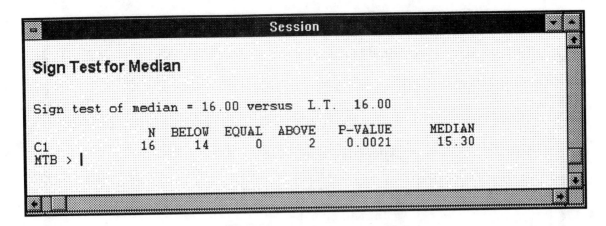

Minitab 15.2

Exercise 1: Work *Example 1* using the classical approach.

2. THE WILCOXON SIGNED-RANK TEST

Note: The *Wilcoxon Signed-Rank Test* as described is also known as the *Wilcoxon Matched-Pairs Signed-Rank Test* since this test can be used for *dependent* samples.

Example 2: Use **MINITAB** to apply the *Wilcoxon Sign-Rank* test to *Example 1*.

To do this, select **Stat→Nonparametrics→1-Sample Wilcoxon** and the **1-Sample Wilcoxon** dialog box will be displayed. Select column C1 for the **Variables** text box and select the check box for the **Test median**. Make sure the value 16 is in the **Test median** text box and **less than** is selected in the **Alternative** drop down box. Figure *Minitab 15.3* shows the **1-Sample Wilcoxon** dialog box with the appropriate entries.

Minitab 15.3

Click on the **OK** button and the **Session** window will display the results. The **Session** window is shown in figure *Minitab 15.4*. Observe that the *p*-value for this test is 0.001 (rather small compared to a significance level of 0.05), so we can reject the null hypothesis and claim that the median weight for the bags of corn chips is less than 16 ounces. This is the same conclusion as when the **Sign** test is used. Since the Wilcoxon Signed-Rank test uses more information than the sign test, it will be a more efficient test. An indication of this is shown by the valued of the *p*-values. For the sign test, the *p*-value is 0.0021 as compared to 0.001 for the Wilcoxon Signed-Rank test.

Minitab 15.4

Note: This version of **MINITAB** does not have a procedure to *directly* do the *Wilcoxon Matched-Pairs Signed-Rank Test*. That is, a test for dependent samples. However, we can manipulate things to enable **MINITAB** to do an equivalent test. First, enter the two dependent sets of values in columns *C1* and *C2*. Find the difference of the values in C1 and C2 and save in column C3. Then apply the **1-Sample Wilcoxon** to the differences in column C3.

Exercise 2: Work *Example 2* from a classical approach.

3. RANK SUM TEST FOR COMPARING TWO POPULATIONS

Note: The *Rank Sum Test for Comparing Two Populations* is also known as the *Mann-Whitney Test*. This test is used for *independent* samples.

Example 3: To determine whether reflex reaction time is a function of age, independent samples of ten 16-year-old females and ten 30-year-old females were selected and tested. The following gives the reaction times in seconds for a certain stimulus.

16-year	4.33	5.03	2.10	3.37	2.85	2.93	4.23	3.4	3.49	4.12
30-year	5.23	3.59	2.75	4.59	5.85	4.88	5.78	6.48	5.12	4.96

Use **MINITAB** to test whether the reaction times are different for the two populations.

To apply the *Mann-Whitney test* to the two sets of *independent* data, enter the 16-year and 30-year values in column *C1* and *C2*, respectively. Next, select **Stat→Nonparametrics→Mann-Whitney** and the **Mann-Whitney** dialog box will be displayed. Select the appropriate entries as shown in figure *Minitab 15.5*.

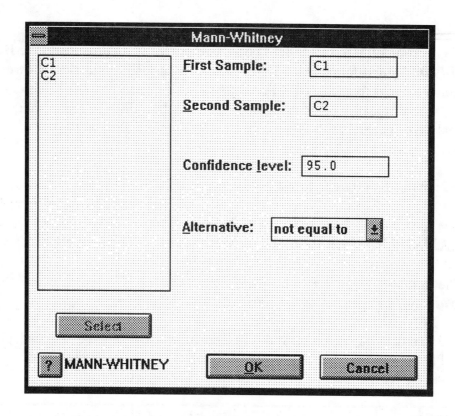

Minitab 15.5

Click on the **OK** button and the associated computations will be displayed in the **Session** window. This is shown in figure *Minitab 15.6*.

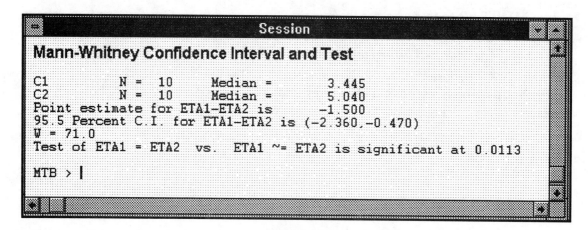

Minitab 15.6

The test statistic value is given by W = 71.0 and the *p*-value = 0.0113. Thus, at the 5% level of significance, we will reject the null hypothesis that the two population medians are identical with respect to reaction to the stimulus, since 0.0113 < 0.05.

Note: Also included in the output are the medians for the two samples, the point estimate for the difference of the two population medians, and the 95% confidence for this difference.

4. THE RUNS TEST FOR RANDOMNESS

Example 4: Two professors, X and Y, are scheduled to teach a section of introductory statistics during the same time period next semester. The Statistics department claims that students are randomly assigned to one of the sections as they register for the course. Forty students signed up for the course, and the section assignments in the order of registration are given below. Here, X represents the section taught by professor X and Y represents the section taught by professor Y.

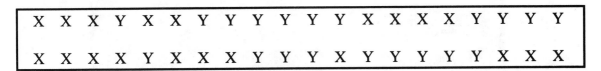

Use **MINITAB** to determine whether there is sufficient evidence at the 5% level of significance to reject the claim that section assignments were made at random.

Enter X as a 1 and Y as a 0 in column C1, for each of the 40 values since X and Y are non-numeric values. Select **Stat→Nonparametrics→Runs Test** and the **Runs Test** dialog box will be displayed. Select C1 for the **Variables** text box and select the check box for **Above and below the mean**. By selecting this option you are telling **MINITAB** to use the mean as the baseline to determine the number of runs. The **Runs Test** dialog box is shown in figure *Minitab 15.7*.

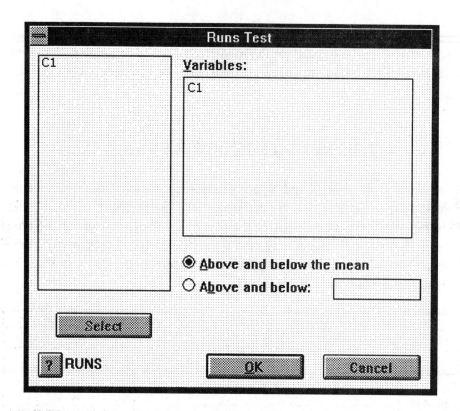

Minitab 15.7

Click on the **OK** button and the **Session** window will display the results. This is shown in figure *Minitab 15.8*. Observe that the *p*-value computed by **MINITAB** is 0.0105, so we can reject the hypothesis that the assignments were made at random, at the 5% level of significance.

Minitab 15.8

LAB #15: *DATA SHEET*

Name: _____   *Date:* _____

Course #: _____   *Instructor:* _____

1. The median age of a population was established to be 32 years in 1990. Suppose that another research team later repeated the procedure on a sample of 18 counties in the population that was studied. The median age for these 18 counties are given in the table below. Using the *Sign* test, could the second team conclude, at the 0.05 level of significance, that the median age in the population was different from 32 years?

Obs.	1	2	3	4	5	6	7	8	9
Median Age	29.6	31.2	32.3	37.0	31.4	32.6	34.5	32.0	30.3

Obs.	10	11	12	13	14	15	16	17	18
Median Age	33.0	35.4	33.4	31.0	32.7	32.0	32.5	35.4	31.6

Use **MINITAB**, when appropriate, to help respond to the following questions.

(a) State an appropriate null hypothesis and alternative hypothesis from this information.

H_0:

H_1:

(b) Give the *equation* of the *test statistic* for this test. Define each symbol that you use in the equation.

(c) What is the value of the *test statistic*?

Test statistic = _____.

(d) What is the *p-value* for the test?

p-value = _____.

(e) What can you *conclude* from this test?

2. AIDS, one of the deadliest diseases ever known to man, is spreading at an incredible rate. As of December 1993, this disease had done most of its damage to Africa and the Americas, with the median number of cases in the countries of Africa to be 884. The following data set lists the number of aids cases in 10 randomly selected countries in the Americas. Can we conclude, at the 0.05 level of significance that the median number of aids cases in the Americas is greater than 884 when using the *Wilcoxon Signed-Rank* test?

| 397 | 15 | 8640 | 389 | 39 | 16091 | 1329 | 525 | 1 | 883 |

Use **MINITAB**, when appropriate, to help respond to the following questions.

(a) State an appropriate null hypothesis and alternative hypothesis from this information.

H₀:

H₁:

(b) Give the *equation* of the *test statistic* for this test. Define each symbol that you use in the equation.

(c) What is the value of the *test statistic*?

Test statistic = _____.

(d) What is the *p-value* for the test?

p-value = _____.

(e) Using a level of significance of $\alpha = 0.05$, what is the *critical value* for this test?

Critical value = _____.

(f) What can you *conclude* from this test?

3. The following table lists the total precipitation (in inches) for each month of 1995, and the departure from normal precipitation for each month for Louisville, Kentucky. Since the data are recorded in order of occurrence and they may be dichotomized according to whether the departure from normal is positive or negative, a one-sample *Runs* test for randomness may be used to determine whether the pattern of departures is the result of a random process.

Month	*Total Precipitation*	*Departure from Normal*	*Runs*
January	3.20	+0.34	1
February	2.00	-1.30	2
March	2.17	-2.49	2
April	2.64	-1.59	2
May	9.51	+4.89	3
June	2.84	-0.62	4
July	3.40	-1.11	4
August	4.07	+0.53	5
September	1.32	-1.84	6
October	5.42	+2.71	7
November	2.37	-1.33	8
December	3.28	-0.36	8

Use **MINITAB**, when appropriate, to help respond to the following questions.

(a) State an appropriate null hypothesis and alternative hypothesis from this information.

   H$_0$:

   H$_1$:

(b) Give the *equation* of the *test statistic* for this test. Define each symbol that you use in the equation.

(c) What is the value of the *test statistic*?

   Test statistic = _____.

(d) What is the *p-value* for the test?

    p-value = _____.

(e) Using a level of significance of $\alpha = 0.05$, what is the *critical value* for this test?

    Critical value = _____.

(f) What can you *conclude* from this test?

4. In the past, The Wright Patterson Air Force Base in Dayton, Ohio has reported the number of cited UFO crashes that contained alien bodies. Since we do not know whether the government has manipulated data concerning UFO crashes, reliable data is suspect. The following table gives a preliminary list of UFO crashes in order from 1939 to 1992.

In the table:

A -- denotes a crash that consisted of any alien life forms that survived
D -- denotes a crash that none of the alien life forms survived the crash

Use a one-sample *Runs* test for randomness to determine whether the survival or non-survival of the alien life forms is the result of a random process. Use a 0.05 level of significance.

Crash Number	A or D	Crash Number	A or D	Crash Number	A or D
1	A	12	D	23	A
2	D	13	D	24	D
3	D	14	D	25	D
4	D	15	D	26	D
5	A	16	D	27	D
6	D	17	D	28	A
7	D	18	A	29	A
8	D	19	D	30	D
9	D	20	D	31	A
10	D	21	A	32	A
11	D	22	A	33	A

Use **MINITAB**, when appropriate, to help respond to the following questions.

(a) State an appropriate null hypothesis and alternative hypothesis from this information.

H₀:

H₁:

(b) Give the *equation* of the *test statistic* for this test. Define each symbol that you use in the equation.

(c) What is the value of the *test statistic*?

Test statistic = _____.

(d) What is the *p-value* for the test?

p-value = _____.

(e) Using a level of significance of $\alpha = 0.05$, what is the *critical value* for this test?

Critical value = _____.

(f) What can you *conclude* from this test?

5. Below is a table of educational attainment for fourteen southern states of persons 25 years and over for the years 1980 and 1990. We would like to determine whether the median percentage of bachelor's degree or higher in 1990 was higher than the median percentage of bachelor's degree or higher in 1980. *Hint: Use the Rank Sum (Mann Whitney) test.*

State	*% Bachelor's degree or higher in 1980*	*% Bachelor's degree or higher in 1990*
Alabama	12.2	15.7
Arkansas	10.8	13.3
Florida	14.9	18.3
Georgia	14.6	19.3
Kentucky	11.1	13.6
Louisiana	13.9	16.1
Mississippi	12.3	14.7
North Carolina	13.2	17.4
Oklahoma	15.1	17.8
South Carolina	13.4	16.6
Tennessee	12.6	16.0
Texas	16.9	20.3
Virginia	19.1	24.5
West Virginia	10.4	12.3

Use **MINITAB**, when appropriate, to help respond to the following questions.

(a) State an appropriate null hypothesis and alternative hypothesis from this information.

   H₀:

   H₁:

(b) Give the *equation* of the *test statistic* for this test. Define each symbol that you use in the equation.

(c) What is the value of the *test statistic*?

   Test statistic = _____.

(d) What is the *p-value* for the test?

   p-value = _____.

(e) Using a level of significance of α = 0.05, what is the *critical value* for this test?

   Critical value = _____.

(f) What can you *conclude* from this test?

6. Below shows some statistics for the margin of the winner (measured in lengths) for the Kentucky Derby during the years 1875 to 1895 and 1975 to 1995. Because of modern training and breeding methods, we would like to determine whether the median length of the winners for the period 1975 to 1995 is larger than that for the period 1875 to 1895. *Hint: Use the Rank Sum (Mann Whitney) test.*

Date	Winner	Margin (lengths)	Date	Winner	Margin (lengths)
May 17, 1875	Aristides	2	May 3, 1975	Foolish Pleasure	1.75
May 15, 1876	Vagrant	2	May 1, 1976	Bold Forbes	1
May 22, 1877	Baden-Baden	2	May 7, 1977	Seattle Slew	1.75
May 21, 1878	Day Star	1	May 6, 1978	Affirmed	1.5
May 20, 1879	Lord Murphy	1	May 5, 1979	Spectular Bid	2.75
May 18, 1880	Fonso	7	May 3, 1980	Genuine Risk	1
May 17, 1881	Hindoo	4	May 2, 1981	Pleasant Colony	0.75
May 16, 1882	Apollo	1.5	May 1, 1982	Gato Del Sol	2.5
May 23, 1883	Leonatus	3	May 7, 1983	Sunny's Halo	2
May 16, 1884	Buchanan	1	May 5, 1984	Swale	3.25
May 14, 1885	Joe Cotton	0.25 (neck)	May 4, 1985	Spend a Buck	5.75
May 14, 1886	Ben Ali	0.5	May 3, 1986	Ferdinand	2.25
May 11, 1887	Montrose	2	May 2, 1987	Alysheba	0.75
May 14, 1888	Macbeth II	1	May 7, 1988	Winning Colors	0.25
May 09, 1889	Spokane	0 (nose)	May 6, 1989	Sunday Silence	2.5
May 14, 1890	Riley	1.75	May 5, 1990	Unbridled	3.5
May 13, 1891	Kingman	0.5	May 4, 1991	Strike the Gold	1.75
May 11, 1892	Azra	0 (nose)	May 2, 1992	Lil E. Tee	1
May 10, 1893	Lookout	4	May 1, 1993	Sea Hero	2.5
May 15, 1894	Chant	6	May 7, 1994	Go for Gin	2
May 06, 1895	Halma	5	May 6, 1995	Thunder Gulch	2.25

Use **MINITAB**, when appropriate, to help respond to the following questions.

(a) State an appropriate null hypothesis and alternative hypothesis from this information.

H_0:

H_1:

(b) Give the *equation* of the *test statistic* for this test. Define each symbol that you use in the equation.

(c) What is the value of the *test statistic*?

   Test statistic = _____.

(d) What is the *p-value* for the test?

   p-value = _____.

(e) Using a level of significance of $\alpha = 0.05$, what is the *critical value* for this test?

   Critical value = _____.

(f) What can you *conclude* from this test?

7. **Team Exploration Project**. Use your library or any other resource to collect a set of data that could be analyzed with the *Runs test*. Use **MINITAB** to do a complete analysis of the data. Present any relevant analysis and discussions in a report. If you select data from published research, you should include relevant factors about the validity and reliability of the study.

8. **Team Exploration Project**. Use **MINITAB** to simulate the rolling of a twelve sided die. Test whether the simulation is random. Present your findings in a report. *Hint: Simulate a large set of values. Let an even number be represented by 0 and an odd number be represented by 1. Test whether these runs are random.*

9. **Team Exploration Project**. Collect data for *any lottery game*. Use **MINITAB** to determine whether the machine that selected the values is fair. Present your findings in a report. *Hint: Collect data as they appear in the drawings. Let an even number be represented by 0 and an odd number be represented by 1. Test whether these runs are random.*

NOTES

APPENDIX A

Lifeline Data

See Lab #8 — Example 1

Observation Number	Age (years)	Length of Lifeline (nearest 0.15 cm)
1	19	9.75
2	40	9.00
3	42	9.60
4	42	9.75
5	47	11.25
6	49	9.45
7	50	11.25
8	54	9.00
9	56	7.95
10	56	12.00
11	56	8.10
12	57	10.20
13	57	8.55
14	58	7.20
15	61	7.95
16	62	8.85
17	62	8.25
18	65	8.85
19	65	9.75
20	66	8.85
21	66	9.15
22	66	10.20
23	67	9.15
24	68	7.95
25	68	8.85
26	68	9.00
27	69	7.80
28	69	10.05
29	70	10.05
30	71	9.15
31	71	9.45
32	71	9.45
33	72	9.45
34	73	8.10
35	74	8.85
36	74	9.60
37	75	6.45
38	75	9.75
39	75	10.20
40	76	6.00
41	77	8.85
42	80	9.00
43	82	9.75
44	82	10.65
45	82	13.20
46	83	7.95
47	86	7.95
48	88	9.15
49	88	9.75
50	94	9.00

APPENDIX B

Fleer Basketball Data

SeeLab #8 -- Exercise 1

Observation Number	PRICE ($)	NYEARS (YEARS)	HEIGHT (FEET)	WEIGHT (POUNDS)	APSPG	ARPG
1	0.18	7	7.33	290	6.5	8.50
2	0.66	5	6.83	215	11.5	5.40
3	0.18	5	6.41	190	13.6	4.10
4	0.39	2	6.25	170	15.3	2.30
5	3.29	5	6.83	250	23.0	12.60
6	1.05	1	6.41	180	12.7	2.80
7	7.89	1	6.41	215	22.7	5.50
8	1.32	4	6.58	215	21.6	4.30
9	0.18	2	6.58	190	10.6	3.10
10	0.39	9	6.91	260	13.6	10.56
11	0.18	3	6.58	190	12.6	2.32
12	4.61	4	7.00	240	23.6	10.10
13	0.18	4	6.75	245	12.5	11.50
14	1.97	7	6.67	200	26.1	7.00
15	4.61	10	6.75	220	24.5	10.00
16	6.58	1	6.58	195	15.7	4.20
17	0.18	3	6.67	220	19.1	6.50
18	0.53	6	6.33	205	20.0	2.70
19	0.53	2	6.83	220	11.5	7.60
20	19.74	5	6.50	200	32.6	6.30
21	3.95	2	6.67	210	14.2	6.00
22	0.92	6	6.58	200	7.9	3.60
23	0.92	11	6.83	225	18.7	7.60
24	0.18	9	6.58	215	18.5	4.70
25	3.29	5	6.08	175	12.5	2.40
26	0.92	3	7.00	230	7.1	4.60
27	1.18	8	6.83	230	21.0	6.80
28	3.29	1	6.50	220	11.4	5.20
29	0.53	7	6.75	235	21.7	8.60
30	3.29	4	6.75	255	25.6	10.80
31	0.18	1	6.92	265	4.3	3.10
32	3.29	5	6.50	255	23.3	11.70
33	1.97	1	6.33	190	12.4	3.50
34	2.63	1	6.25	175	10.8	2.90
35	0.66	2	6.75	210	13.8	5.40
36	3.29	1	6.33	185	16.7	2.80
37	0.18	7	6.33	220	15.6	2.50
38	1.05	7	6.75	225	18.6	5.60
39	0.53	5	6.33	195	16.3	3.10
40	4.61	10	6.75	220	19.7	7.30
41	0.18	5	6.75	235	14.2	7.90
42	7.89	1	6.83	230	19.2	7.20
43	0.18	2	7.00	240	5.3	4.30
44	1.58	1	7.33	250	12.7	6.70
45	6.58	1	6.50	190	11.1	4.60

NOTES

NOTES